SPACE SCIENCE, EXPLORATION AND POLICIES

THE DESTINIES OF THE STARS

SPACE SCIENCE, EXPLORATION AND POLICIES

Additional books and e-books in this series can be found on Nova's website under the Series tab.

SPACE SCIENCE, EXPLORATION AND POLICIES

THE DESTINIES OF THE STARS

SVANTE ARRHENIUS

JOENS ELIAS FRIES
(TRANSLATOR)

Library of Congress Cataloging-in-Publication Data

ISBN: 978-1-53614-956-2

Published by Nova Science Publishers, Inc. † New York

CONTENTS

PREFACE*

Since I presented *Worlds in the Making* and *Life of the Universe as Conceived by Man from Earliest Ages to the Present Time* to the public—which received them with far greater interest and appreciation than I could foresee—I have had repeated occasions to treat new questions of a cosmological nature, questions largely arisen from new discoveries and observations within the scope of astronomy. Vast new vistas have been opened through the study of the relation of the stars to the "Milky Way" and through observations of our neighbor planets. The last mentioned give plain indications of the course of planetary evolution and thus enable us to surmise the changing fate and future position of the Earth. In an earlier German publication, *Das Schicksal der Planeten* (1911), I dealt with some of these problems. As, further, the evolution of the solar system from the Milky Way nebula, to which I have devoted several lectures at home and abroad, may be considered as the pre-history of the evolution of the planets, I have given this collection of cosmogenic articles the common title *The Destinies of the Stars*. I offer as introduction a lecture delivered before the Fourth International Philosophical Congress in Bologna, 1911, dealing with the "Origin of Star-Worship."

---

* This is an edited, reformatted and augmented version of *The Destinies of the Stars* by Svante Arrehnius, originally published by G. P. Putnam's Sons, dated 1918. The views, opinions, and nomenclature expressed in this book are those of the authors and do not reflect the views of Nova Science Publishers, Inc.

Hoping that this little book will, to a considerable extent, fill the gaps in my previous works, I present these treatises in remodelled form.

Svante Arrhenius
Stockholm, *November, 1915*

Translator's Preface

When Dr. Svante Arrhenius in the year 1903 received the Nobel Prize in Chemistry it came as a fitting reward of his achievements principally in the electro-chemical field. It was natural, however, that a genius of his calibre would not limit his interest to the "infinitely small" but would gradually broaden it to encompass the "infinitely large." And "to take an interest" means with Dr. Arrhenius to push the boundaries of the unknown and of the unexplored a little farther away from man. His evolution in this respect runs parallel with that of all the great men who stand out as leaders in the history of science. Wrapt up in the solution of a particular problem and fired with the divine yearning to reach ultimate causes they are inevitably driven to ever widening circles of research until this whole material universe, its whence and whither, becomes the overpowering passion of their spirits. Thus the mere titles of the works of Dr. Arrhenius, read in the sequence of their publication, give us, better than any biography, the history of a soul, which, no matter what his unprofessed philosophical faith may be, constitutes our strongest evidence in favour of that theory of "purposiveness" in the universe which Dr. Arrhenius so heartily abhors (and justly so) when resorted to in natural science, but which theory nevertheless (and justly so) is so dear to the philosopher:—*Researches in Regard to the Conductivity of Electrolytes; Conductivity of Extremely Diluted Solutions; Chemical Theory of Electrolytes; Textbook in*

Theoretical Electro-Chemistry; Textbook in Cosmological Physics; Worlds in the Making; Infinity of the Universe; Life of the Universe as Conceived by Man from Earliest Ages to the Present Time;—thus run the titles of a few of the works we already have by Dr. Arrhenius' hand. How were it possible for him NOT to write *The Destinies of the Stars*? The volume came as inevitably as fruition follows flowering. What remains to be seen is if Dr. Arrhenius can withstand the tremendous temptation that must be at work in his soul to lift, be it ever so little, the curtain that separates natural science and philosophy; we hope he will give in; we admire in this book how he reads "The Riddle of the Milky Way"; we certainly wish to know how he reads—the riddle of the universe.

The Destinies of the Stars met with unexampled success in Sweden. Three editions appeared within two months when the book was published in November, 1915. The American version has been somewhat delayed principally due to war conditions. This, however, has not been wholly a loss to the English-speaking world as Dr. Arrhenius by no means has been idle in the meantime. Considerable additional subject matter, including three new pictures, has been added, chiefly based on the most recent astronomical discoveries some of which have been recorded as late as 1917.

For valuable suggestions and for all the American equivalents of the metric measures in the original, the reader as well as the translator, is indebted to a member of the Publishing House that presents this volume in such an attractive way, Mr. E. W. Putnam, himself an ardent lover of astronomy and a writer on the subject.

Dr. Arrhenius is justly renowned for his lucid style and polished form. All that is lacking in these qualities within the covers of this volume is wholly due to the deficiencies of the translator, who however could not resist the temptation of transferring to Anglo-Saxon soil this monument to the genius of his former teacher, Dr. Svante Arrhenius.

J. E. Fries

Chapter 1

The Origin of Star-Worship

Astronomy occupies a rather unique position among the natural sciences. While physics, chemistry, and the biological sciences form the foundation of the extraordinary material development of our day, astronomy, in the eyes of most people, is of little practical value. What benefit could we derive from knowing whether a star lies a hundred or a thousand billion miles from the Sun, or from understanding how the stellar bodies have evolved in the course of billions of years? And yet astronomy has not been as futile as is commonly imagined, neither is it useless at the present time. This science is still of the greatest importance to common life by fixing our standards of time and, before the introduction of the compass, was invaluable also to navigation, which art, moreover, depends now upon astronomy for determination of geographical position on the open sea. Observations for these purposes, however, are of such a simple nature, that they hardly fall under the astronomical science proper, but rather under the applied sciences. They have entered into daily life much as, in commerce, the determination of the weight of a body is not considered as belonging to the science of physics, although it depends on the use of a physical instrument, the scale.

But we must not forget that what we now consider so commonplace that it entirely has lost the grand aspect of science, once was the goal of groping scientific endeavour. All natural science has grown out of the needs of practical life.

Geometry is probably even older than astronomy. The name means: to measure land, and the oldest geometry was, accordingly, devoted to the measurement of distances on the Earth and later to the determination of the area of land-holdings. This extremely important practical application of geometry is of such a simple nature that it is not mentioned in modern mathematical science, to which geometry belongs. In this manner, the original theses of all our natural sciences have become the possession of the public to such a degree that they are looked upon as self-evident. This is the case also with those parts of astronomy which, because of their practical importance at the outset, gave rise to the science itself.

The growing knowledge about the stars, like all higher insight, became among primitive peoples the private possession of their leaders, was by these kept a secret and made a part of the venerable realm of religion. We find that a majority of these old peoples rendered worship to the stars, believing them to govern the fates of human beings. This may indeed seem highly remarkable, as our everyday experience is that the stellar bodies, with the exception of the Sun, exert no perceptible influence on organic nature, and such conclusion is emphatically confirmed by the systematic collection of all our experiences which we call modern science. The Sun, as stated, is the exception as it reigns over the entire nature, the living as well as the lifeless, by virtue of the heat and light which abundantly flow from this autocrat of our planetary system. Perhaps the Moon also plays some small active part as it seems somewhat to affect the barometric pressure, the magnetic and particularly the electric conditions of the atmosphere, which, in turn, appear to influence several life processes. On the other hand we cannot point to any influence upon nature traceable to the other stellar bodies.

Obviously, primitive man devoted his thoughts only to such conditions as affected his interests in a beneficial or detrimental way. On the assumption that these conditions were governed by spirits who resembled man and who in particular were endowed with will, our ancestors endeavoured by sacrifice and exorcism to move the spirits they feared to discontinue their harmful activity. Some such spirits dwelt in beasts of prey and in other noxious animals, such as poisonous snakes; others caused earthquakes, volcanic eruptions, snowstorms, lightning ravages, fires, floods, heat, drought, etc. Against these calamities religious exercises formed a protection. Religion sprang mainly from fear of spirits. Later on thank offerings and hymns were bestowed upon benevolent objects and phenomena in nature.

It is evident that this primitive, simple religion is of far greater age than star-worship. The latter presupposes a comparatively high degree of culture. The stars were of no benefit to man until it became necessary to measure time intervals comprising a greater number of days than might be counted on one's fingers. How this growth in all probability took place we shall endeavour to explain in the following. Fairly certain it is that star-worship did not grow out of man's admiration for the sublime drama which at dawn of morning commences at the eastern horizon, and proceeds in the course of a day over the firmament in order to close before night just beyond the western expanses, neither was it founded on man's gratitude toward the torch-bearers of night for their incessant battle against gloomy clouds and all the other spirits of darkness.

Even tribes on a rather low stage have no doubt noticed the most conspicuous among the stellar bodies. The attitude of the Australian aborigines is significant in this connection. According to Spencer and Gillen they possess legends about the Moon, which is male, about the Sun and Venus, about the pernicious Magellanic clouds, and about the Pleiades, who, like the Sun and Venus, are considered female. Eclipses have, as is natural, attracted the greatest attention. While these primitive men indulge in an incredible number of religious ceremonies pertaining to conditions of their daily life, none of these exercises are devoted to the stars, if we except stone throwing against the Sun during eclipses. Even this treatment is left with considerable serenity of mind for the medicine-men to perform. It is very significant that all stellar bodies are of earthly origin, that the Moon is of male gender while the Sun, Venus, and the Pleiades are female, which indicates that the Moon is considered of greatest distinction. They count time according to "sleeps," i.e., the number of times they have slept, or according to moons; longer periods according to seasons; they have names for summer and winter. They can count to five, or perhaps rather to four, as the term five also means "many." Ideas of power centred in the stellar bodies are evidently absent and therefore also religious ceremonies pertaining thereto. A few tales exist relating to the stars as to other objects within their observation. Thus conditions would undoubtedly have remained but for the high value which the want of a chronometry gave to the regularly changing light of the stellar bodies.

The difference between day and night is of such a deeply fundamental importance that it has left its stamp on the whole organic nature on the surface of the Earth. Plants entirely change their life processes in the course of twenty-four hours; during the day they add to their growth under the stimulus of light; during the night they partly expend the strength gathered in daytime. This cycle is so regular that it functions automatically. The prominent botanist, Pfeffer, has experimented with a Mimosa, which, as is well known, unfolds its leaves during the day but curls them during the night. If left in a dark room any arbitrary day, it nevertheless uncurls its leaves. By means of electric light Pfeffer changed night into day for the plant kept in the dark room. It took considerable time before the plant adjusted itself to the new conditions so that it unfolded its leaves to the influence of the electric light. Animals behave similarly. The night and day period is instilled in their blood. In a certain sense they possess an instinctive chronometry.

It is often stated that the assurance of the return of sunlight after the darkness of night enabled humanity with greater equanimity to acquiesce in the loss of light during one half of its existence and that worship was rendered to the Sun in gratitude therefor. "A new outlook upon life," says Troels Lund, "awakens the moment the great discovery is made that the night of sleep and the night of fear are equally long and always followed by morning and subsequent day." This discovery, however, our predecessors made long before they reached the human stage. Sun-worship by no means derived its origin therefrom.

Rather it is traceable to evidence of the Sun's connection with the changing seasons, although this change also is of domineering influence in the vegetable world inasmuch as the plants store reserve nourishment in the autumn, particularly, and on a large scale, during fructification. Also, lower and higher animals, for example, bees and squirrels, gather winter-stores. It is therefore small wonder if men on a comparatively low stage lay in provisions for the recurring periods of scant food supply.

But a true chronometry beyond five days is foreign to the Australian negro as long as he can only count to four or five. He is aware that Moon-phases reappear and that summer and winter return, but he has no conception of the duration of the time passed between the recurrences of these phenomena. Further progress was made by the people who took the bold step to count the fingers, not on one hand alone, but on both hands, and thus reached the number ten. This was utilized in reckoning time so that the larger unit became a decade, *i. e.* ten days and nights. This unit was original with the Indo-Europeans, Semites, Hindus, Egyptians, and the islanders in the Pacific Ocean. Another advance yet was made in Mexico where the number twenty was introduced corresponding to the sum of all fingers and toes, and thus a unit of time was obtained comprising twenty days and nights. But to rise from these units to one of 365 days was a step exceedingly difficult for the primitive peoples to take.

Thousands of years elapsed before the most intelligent among the races established the length of the solar year. Those who lived in regions where the Sun's altitude notably changes, *i.e.* far from the equator, undoubtedly reckoned time in years, without knowing its length expressed in days. Imagine a nomadic tribe like the Lapps in the north of Sweden. In the autumn their reindeer wander down toward the coast in search of food and the Lapps go with them. In spring the reindeer lead their masters back again to the mountains. It cannot very well escape the observation of these nomads that the Sun shines almost continuously during their stay in the uplands while dreary night reigns nearly without interruption during their sojourn in the lowlands. They are obviously forced to co-ordinate the beautiful summer with the duration of sunlight. To them, therefore, the Sun's extraordinary great importance to life is unquestionable. The same holds true about all people who live sufficiently far from the equator. As a consequence, they become Sun-worshippers. It is not difficult to find examples of peoples who have worshipped the Sun; only a few of the more important ones will be mentioned here.

The people of the bronze age here in the North were zealous Sun-worshippers thousands of years ago, as the many relics from this period, and particularly the rock-carvings, bear witness. The Celts of Western Europe have frequently symbolized the Sun as a cross, while the worship of Moon and stars seems to have been foreign to them as well as to the bronze-age people of the North. The Jewish Samson (Simson) was a Sun-hero, the name being related to the Babylonian Shamash, the Sun-god. In Hesiodos' cosmogony the Sun (Helios) is mentioned before the Moon (Selene). The old Germans worshipped both Sun and Moon, particularly the former. The Slavs possessed a Sun-god Dazbogu, but no deity identified with the Moon. Similar conditions obtained among the Finnish forefathers. The Chinese Tao-priests light fires during the vernal equinox as we do at Walpurgis and midsummer, and they sacrifice rice and salt to the flames. "This is a remnant of the Sun-cult," says Solomon Reinach, from whom these data in regard to Sun-worship principally are taken. In Japan, the Moon is of male sex, the Sun of female sex, which indicates that there, as with the Australian negroes, the Moon was originally considered more important than the Sun. Nevertheless the Japanese are now Sun-worshippers; the Sun is placed as emblem of the most high in their flag and the Mikado is known to trace his lineage from the Sun. They have, therefore, long ago passed from Moon-cult to Sun-cult. It is probable that this step was taken even earlier in China, where the Sun furthermore is of male sex. With growing civilization all people learn to understand, as have the Japanese, the superior importance of the Sun. The Incas of Peru, who reached a very high grade of culture, were sun-worshippers and called themselves children of the Sun, although they lived near the equator where the Moon-cult, as we presently shall see, owns its most faithful adherents.

In the neighbourhood of the equator, winter and summer differ very slightly with respect to temperature and altitude of the Sun. Rather, it is the alternation of humid and dry seasons that is of deciding importance. No sheet of snow covers the ground in winter-time, kills the vegetation, or decimates the supply of nourishment for animals and men. Indeed, contrary to our experience, a suppression of growth often accompanies a high altitude of the Sun due to the drought which simultaneously occurs. The altitude and luminosity of the Sun change altogether too slightly in the course of the year to attract the attention of primitive man. The light of the Moon, on the other hand, varies from full intensity to nothing and this takes place in periods so short that memory has no time to forget the cycle. Even the low-standing Australian negroes utilize the phases of the Moon to denote remote times. Chronometry in any true sense they do not possess, unable as they are to state the number of days in a month. How much more fortunate the peoples who could count to ten or better yet to twenty and thus were able to use the single or the double decade as a measure of time. For them it was easy to determine the time between two successive phases of the Moon, seven and a half days apart.

The truth once grasped that four phases separate two new Moons the important bridge could be established between the short measure, a day, and the longer measure, a month. The latter was then found to be nearly thirty times longer than the former. On a higher stage of culture, it was discovered that this ratio was not exact and the discrepancy must greatly have puzzled the people. The correct ratio is 29.53. At all events the periodic return of full Moon and new Moon proved the most reliable measure of time within their experience. This was something entirely different from the irregular occurrences of earthquakes, storms, lightning, and floods, not to mention the ravages of beasts and foes. Spaces of time that hitherto appeared boundless could be surveyed and computed. The idea of eternity dawned for the first time on humanity. The Moon was the great master, measurer of all. In Sanskrit the Moon's name is *Mâs*, *i. e.* measurer, and "*mensis*" (Lat. month) is closely related to "*mensura*" (Lat. measure).

With peoples who did not live too far from the equator, the Moon, therefore, took precedence of the Sun. Among the Mexicans existed, long ago, a peculiar unit for measuring long periods of time, called "*tonalamatl*," comprising 260 days. It was undoubtedly intended to contain nine synodical months (reckoned from new Moon to new Moon). But such a period would consist of 265.58 days and so could not be made to agree with the double decade and was therefore shortened to 260 days, as we round off the solar year, in reality 365.24 days, to an even 365 days. Elaborate studies have been made in order to explain why the Mexicans chose nine synodical months instead of twelve, as most other peoples, but the question has not been solved. This much is certain, that "*tonalamatl*" has nothing to do with the solar year, but only with the month. The high age of "*tonalamatl*" is proved by the fact that the priests adhered to it in magic and horoscope casting long after the solar year came into public use. A learned Mexican, de Jonghe, has pointed out that "*tonalamatl*" was used by all the tribes belonging to the Nahua-group, which tribes separated very long ago. This unit of time, therefore, bears all the evidence of a very high age, but is obviously younger than the synodical month.

Our information in regard to Moon-worship among the peoples of Mesopotamia is even more explicit. The Moon (Sin) was rendered homage far earlier than the Sun (Shamash). The following translation of a hymn in cuneiform letters I quote from L. Bergström, who published a study on "Semitic Moon-Worship" in *Nordisk Tidskrift*, 1909:

Oh, Sin, thou who alone givest light,
Thou, who bringest light to men,
Thou, who showest favour to the dark-tressed ones,
Thy light shines on the firmament,
Thy torch illuminates like fire,
Thy radiance fills the wide earth.
Oh, heavenly Anu, whose insight and wisdom no one comprehends,
Thy light is splendid as Shamash, thy firstborn,
Before thee prostrate the great gods themselves in the dust
For on thee rests the fate of the world.

Anu was god of the heaven and seems here to stand for god in general. Sin was father to the daughter Shamash, who in this hymn already is considered almost comparable to the father. Later on during the Hammurabi-dynasty (about 2000 B.C.) the Sun, Shamash, was accepted as supreme god, but the Moon remained the regulator of time for religious purposes. For astrological forecasts the priests preferred to use the Moon, and the "signs" in the Moon were the most important. This is true of astrological prophecies also at the time of Tycho Brahe. "Oh, Sin, thou tellest the oracles to the gods who pray them of thee," reads an incantation. From Babylon, the heart of civilization, Moon-worship spread to Arabs and other Semites, and with the Hebrews, as Bergström remarks, the Moon originally played a far greater part than the Sun, although at the time of Christ the condition was reversed. Nevertheless, the Moon has still retained its position as chronographer in the Church Calendar. In Psalms 104:19 we read: "He appointed the Moon for seasons."

In general, we have built on the belief that the stellar bodies made an impression on men by virtue of the light they radiate and it has therefore been difficult to explain why the Moon ranked above the Sun. It is frequently said that the Sun (in Babylonia) was considered an enemy of mankind as its heat destroyed the vegetation. (It is true that a period of severe drought occurs there in the height of the summer.) As opposed to this, the moonlit nights would be considered refreshing and salutary. Another explanation is essayed by Bergström, who says that the luminous Moon with its ever-changing shape appealed to the imagination of primitive people in a far higher degree than the fairly constantly brilliant Sun. He is partly correct; the enormous variation in luminosity from full to new Moon enables the observer to notice the difference from night to night. The change by about one hour in each twenty-four hours of the time for the rising and setting of the Moon as compared to the disappearance and reappearance of sunlight, and more important yet the short periods of the Moon phases which leave the observations from each receding phase fresh in memory,—both these phenomena contribute toward the high value of the Moon's synodic revolution as a measure of protracted time duration. The purely practical need dominates, not, probably, any desire to visualize the changes in legends. The Australian negroes use four different names for the Moon according to its four quarters, showing that they, in all probability, believe themselves confronted with four different stellar bodies, just as the Greeks at the time of Homer considered the morning- and the evening-star, *i. e.* Venus, as two separate planets.

On the other hand, no ground worth mentioning supports the theory that the scorching heat of the Sun diminished the peoples' inclination for its worship. On the contrary, homage was generally rendered to phenomena one feared. It is further not true that the Babylonians themselves considered the Sun, Shamash, hostile, the Moon, Sin, friendly. The Sun-god, Shamash, by virtue of his light, was believed to give life and health. The scorching quality of the Sun was attributed to another god, Nergal, prince of the underworld, demon of war and slaughter, source of fever, and, pre-eminently, of plague. No reason existed, therefore, why Shamash should take second rank after Sin, who is said "to carry water and fire," meaning, according to Schrader, that he brought fits of ague and fever. When the Sun after the oppressive day sinks behind the horizon, it is well known that a sharp fall of temperature occurs, particularly in arid zones, but also in humid regions within the tropics where this very phenomenon is utilized for the production of ice. He who exposes himself to the sudden cold of the night falls an easy prey to illness. Particularly is this true under a clear sky—primitive people say "when the Moon shines"—because of the strong radiation. Those who sleep in moonlight are struck by delirium and madness according to primitive thought, an opinion by no means dead among civilized nations—it is common with seafaring men—and is no doubt the origin of the expression: moonstruck (German: *mondsuchtig*, Ital.: *lunatico*, French: *lunatique*, Swedish: *månadsrasande*, etc.). To this belief has probably contributed the fact that epileptic fits frequently possess a period nearly corresponding to the synodical month, which, as I have shown elsewhere, most likely depends on a periodic change in the atmospheric electricity.

In this connection it might also be stated that the third among the great star-deities, Ishtar, the queen of heaven (Astarte, Venus), was the mild but potent, all-merciful sister in every affliction, who delivered from sorcery and illness and gave pardon for sin and guilt. This radiant goddess, who corresponds to the attractive figure of the Catholics' Virgin Mary, was, in spite of her solicitude for the afflictions of men, placed third in the illustrious triad, Sin, Shamash, and Ishtar.

The traveller in the desert is certainly often tortured by the severe drought and a consequent insufferable thirst. This, however, was justly ascribed to the scarcity of water and not to the Sun. The Egyptians, therefore, wished that their departed ones on the journey to their new dwelling-places would meet with refreshing springs where they might quench their thirst and northerly winds that might cool the air. It is well known that the Mohammedans have formed similar conceptions of a paradise in life to come.

Entirely new conditions arose when the population grew until agriculture became a necessity for the production of sufficient food. The influence of the Sun now became so dominating that it must be given the first place among the powers that affect the fates of men. The plants have a decided annual period and so have the overflows of the rivers which were of the utmost importance in those countries where the cradle of civilization stood. The rainfalls themselves were of annual recurrence as were also the intervening droughts. In Egypt the great importance of the floods caused the introduction of the solar year at a very early stage and its length was fixed at 12 months of 30 days, or 360 days, so that the beginning of the year occasionally had to be adjusted. This was accomplished by resorting to observations of the rising and setting of the Dog Star or Sirius. Thus we realize how difficult it was to determine the exact length of the solar year from everyday phenomena. The great reformer Amenhotep IV. endeavoured about 1400 B.C. to have the Sun-god recognized as sole master of all the world. He met, however, with such great obstacles on the part of the conservative priesthood, which as they largely were serving different gods would have lost part of their power if the reform had prevailed, that his successor was obliged to yield to the solid opposition.

In Babylon the local god Marduk, once representative of the planet Jupiter, and among the star-gods ranking next to the three Super-gods, became about 2000 B.C. elevated to the highest position among the gods and assumed at the same time the dignity of Sun-god. It may here be mentioned that Marduk also played a great part as healer of illness. The evolution in ancient Rome followed the same course although at a far later time. Emperor Aurelianus (270–275 A.D.) under the influence of oriental

Mithras-cult elevated the Sun-god to supreme god of the whole Roman Empire, which then comprised almost the entire known world.

Especially significant is the fact that Venus with the Mexicans played as important a part as did the Moon and the Sun. The luminosity of Venus, unlike that of planets outside of the earth's orbit, but like that of the Moon, changes from a maximum, intense enough in the tropics to throw a shadow, down to a minimum which approaches complete darkness. Its period is closely 1.6 years. It falls short of this figure by about two hours and the Mexicans therefore introduced a correction similar to the bissextile day in our leap year, of one day in every twelve years, which day, however, had to be deducted instead of added. Observations of the changing luminosity of Venus and of its position relative to the Sun obviously lent themselves admirably to measurements of long periods of time and particularly to determination of the length of the important solar year as five Venus-periods very nearly equal eight solar years. The Mexican priests established the fact that 104 solar years correspond to 65 Venus-periods or 146 "*Tonalamatl.*"

Star-cult was as strongly developed in Mexico as in Babylon. Its main doctrine is stated by Alfredo Chavero thus: "The Father-Creator was Heaven, Xiuhtecutli, or the Azure-blue master. The mother was Omecihuatl, the Milky Way, or the dual mistress." It is well known that a large part of the Milky Way, from the "Swan" to the neighbourhood of the Southern Cross, is divided in two parallel branches, which fact probably is responsible for the title, the "Dual Mistress." "Heaven influenced the Milky Way through fire; from its cosmic matter the stars were set free, the most prominent of which were Tonatiuh, the Sun; Tezcatlipoca, the Moon; and Quetzalcoatl, Venus. These were made the supreme gods. For the purpose of worship, they were symbolized in human shape. Myriads of images, representing these star-gods, were modelled in clay, wood, or stone."

According to this remarkable picture the Mexicans should have arrived at a far better solution of the world-riddle than even the Babylonians did. While most other peoples assumed heaven and earth as the original principles, they gave the high position of progenitress of all to the Milky

Way. From her the innumerable stars, with the Sun in the lead, issued. This agrees to a considerable extent with the present conception which we have arrived at during this very last decade, principally thanks to the work of American astronomers. Their investigations have shown how the stars are segregated from the nebulous primeval matter of the Milky Way; how they add distance between themselves and their matrix with age, while simultaneously they develop an increasing individuality.

We have seen that Venus-Ishtar was honoured with membership in the august triad of star-gods, also among the Babylonians. Their successors, the Assyrians, retained the inherited traditions. Thus their kings in the ninth century before Christ symbolized their divine lineage by wearing a necklace with a moon-crescent in the middle, a cross in a ring, emblem of the Sun, on one side, and on the other a star, emblem of Venus (compare Montelius, *Nordisk Tidskrift*, 1904, p. 13, Figure 30). The Jewish synagogues are generally decorated with the star-emblem.

The Mohammedans, like the Jews, utilize the position of the Moon for determination of the Church calendar, and we employ the same means for fixing Easter time. The Mohammedans reckon with a year of 12 synodical months. Twelve such months contain only 354.4 days while a solar year comprises 365.24 days, and as a rule, therefore, the synodical month was rounded to 30 days instead of 29.53 and the solar year to 360 days. Such was the arrangement in Egypt and originally also in Babylon. Primitive men comprehend fractions only with extreme difficulty. In order to correct the discrepancies, odd months were introduced about every sixth year.

From this time we may trace the high reputation of the number twelve. The Zodiac was divided into twelve houses in each of which the Sun was to dwell a month at a time. Day and night were each divided into twelve hours. The circle was divided into 360 degrees corresponding to the number of days of the year, so that the position of the Sun at noon should proceed one degree of the heaven from day to day. As the Moon dominated chronometry, a complication which must have caused considerable confusion was in many places introduced. We have seen that the Australian negroes gave four different names to the Moon in its four different phases. The great change in appearance of the Moon from quarter to quarter makes such a division natural. The synodical month was therefore made to consist of four parts, called weeks. As the length of a month is 29.53 days, the nearest number evenly divisible by four, namely 28, was substituted, and so seven days were allotted to each week, thus introducing an error of not less than 5.5%.

To the establishment of this week the assumption of seven wandering stars has no doubt largely contributed. The priests had discovered that besides Sun, Moon, and Venus, four other stars shift their position on the firmament with reference to the fixed stars, which latter appear always to maintain their relative distances. These four wandering stars were Mercury, Mars, Jupiter, and Saturn. Each day in the week was dedicated to one of the seven wandering stars and received its name. These names have been maintained to the present day, for instance, Sunday, the Sun's day, Monday, the Moon's day, etc. The lunar calendar, established through religious considerations, supplanted the more rational one, which latter, however, survived in Egypt, and was reinstated in the Occident during the French revolution, although unfortunately only for a short time (1793–1805). As a result, the synodical month, in order to suit the calendar, has been changed not only with half a day to thirty days, but also with one and a half days to twenty-eight days. If decades had been adhered to we would have had months of even thirty days and either five leap months of thirty-one days each (during leap year six) or half a decade interpolated at new year.

Besides the seven wandering stars known to antiquity (at present over eight hundred planets have been observed), several other stars and constellations played an important part. The Magellanic clouds, considered of evil nature, and the Pleiades appealed already to the Australian negroes. In the northern hemisphere, where the opportunity of observing the Magellanic clouds is small, situated as they are near the South Pole, the Pleiades have attracted the greater attention and the Phœnicians especially appear to have taken interest in this constellation. From them, reverence for the Pleiades spread to a large part of Africa, where we now to our surprise find this star cluster reproduced along with symbols of Sun, Moon, and Venus. Homer also mentions the Pleiades and a few other constellations, namely, the Hyades, Orion, the Great Bear, and the stars Sirius and Arcturus. At all events, the Pleiades have frequently occupied a unique position in the old world. Sirius, the brightest star in the heavens, and Canopus, the second in brightness, also belonging to the southern hemisphere but only half as far removed from the South Pole as Sirius, have both evoked the attention and the worship of the primitive people, in this case the South Africans.

At length, the nations, particularly the Babylonians and the Mexicans, acquired a wider knowledge of the different stars. As the most important ones, Sun, Moon, and Venus, guided the seasons and hence all natural phenomena, a certain mundane significance was naturally ascribed also to the younger ones. Not only seasons, months, days, and hours were each ruled by its star, but so was everything in nature; different winds, provinces, trades, bodily organs, animals, persons, each possessed its star and celestial protector. Comprehensive studies of these correspondences and connections were made and the conclusions were drawn from immaterial semblances and often wholly arbitrarily,—as regards persons from the configuration of the stars at the time of birth. Thus grew an enormously extensive collection of correspondence- and sympathy-doctrines accompanied by a detailed symbolism, an entire quasi-science, which must never be questioned as it originated with the infallible priests. With the Babylonians, religion and science completely melted together and even art was entirely subservient to the same interests. Occasionally the loss of this blissful state draws forth a sigh. Fortunately it is gone never to return.

The oriental wisdom was brought over to Greece and was there amalgamated with the Platonic-Aristotelian philosophy. In this form the Babylonian heritage held sway over the thought of mankind up to less than 200 years ago. The most important branches of this fanciful, so-called science were astrology and alchemy. Tycho Brahe himself made it the object of his life to strengthen astrology by contributing new material to it. Kepler is said not to have believed in astrology but he nevertheless cast horoscopes not only for princes and persons of high position in order to improve his economy, but also for his own family. Probably traces of the old superstition clung to him, and presumably he thought: "If it does no good, neither does it do any harm."

In the same manner, alchemy was carried on by faithful adepts, but more often by impostors, seldom averse to "occult" sciences. Astrologers and alchemists exist even yet among the numerous devotees to occultism; at high price many of them make their predictions or sell their secrets. I have heard a Swedish engineer of very high standing state that their prognostications agreed with events. Among the few alchemists in Europe, most of whom seem to be religious visionaries, Strindberg is of a certain interest to us. Correspondence-theory has played a very large part in the speculations of the learned up to comparatively recent time. It is utilized extensively in the later fantastic writings of Swedenborg. Numerous traces are to be found also in the weakest works of Strindberg.

The renowned French chemist, Berthelot, has given a valuable analysis of the alchemist's method of treating chemical phenomena. His general conclusion is that the false principles which led the alchemists astray revert back to Plato's and Aristotle's philosophical theories regarding the composition of matter. Something similar can be said of astrology. It plays with ideas of its own fabrication with hardly any foundation in reality. The result is almost wholly without value.

The greatest astronomer in Babylon, Kidinnu (about 200 B.C.), constructed tables of great accuracy giving the position of the stars. In this work he utilized observations gathered over thousands of years. These ephemerides were also intended as scripture source for reading the fate of men and for determination of the auspicious moment for the commencement of an undertaking. At all events, they placed great revenue and power over souls in the hands of the ruling priesthood. It does not appear that these priests were able to rise to an attempt of a physical explanation as to the nature of the stellar bodies. That was probably also considered dangerous. The stars were deities composed of purer and more refined matter than found on Earth. It were not improbable that the gods would inflict vengeance on the presumptuous one who dared to intrude upon their secrets and pass judgment on their peculiarities.

Fortunately, there existed in Greece another tendency in philosophy besides the scholastic and the Platonic-Aristotelian. But this was mainly represented in southern Italy, Sicily, and later in Alexandria. Already the followers of Pythagoras had made important progress toward a solution of the stellar problems. The crowning point was reached by Aristarchos from Samos, who lived in Alexandria about 2100 years ago. He established 1700 years before Copernicus the heliocentric system. It is often said that his work was of little value, as Copernicus nevertheless must do it over again. It is then forgotten that Copernicus himself cites the philosophers of antiquity who expressed theories in agreement with the heliocentric system and expressly states that he was bold enough to advance his hypotheses because so many prominent authorities could be mentioned who favoured them. Copernicus did not dare entirely to break away from the Ptolemaic system, and was inconsistent enough partly to use it in his calculations of the motions of the stars.

We have lately advanced farther along the road of Pythagoras and Aristarchos, of Copernicus and Galileo, and we have perfected their methods to a high degree. Progress in astronomy and kindred sciences is nowadays made at a dizzying speed if measured with the standard of antiquity. Occasionally we hear a warning voice asking us to show more deference to a philosophy directly descending from the Platonic-Aristotelian. He who is at all familiar with the history of natural science will understand us when we answer: “We have had more than enough thereof.”

That non-naturalists sometimes have a peculiar conception of the present status of astronomy is well illustrated by the statement of one of our foremost theologians in a review of a popular astronomical work where he remarked that the astronomer of today had not advanced much beyond those of ancient time who also could forecast eclipses of the Sun. The predictions were then founded on the recurrences of eclipses after regular intervals much as the new Moons were foretold, with the difference only that the latter occur much more frequently.

Our knowledge of the stellar bodies at present and fifty or sixty years ago are a world apart and the same is true of the latter and that of antiquity.

But we must not therefore forget that our brilliant star-science today is derived from men's desire to measure time, and particularly from their need to foresee the food supply in coming days.

Chapter 2

THE RIDDLE OF THE MILKY WAY

During dark but starlit nights, the gorgeous firmament is decorated with an irregular band of light that describes a winding path across the heavens. It continues also in the quarters hidden from our sight so that it may be said to surround the firmament like a girdle. This band, which is most luminous in the Northern Hemisphere, is called "The Milky Way."[1] It forms an angle with the equator of about 60° and divides the firmament in two nearly equal parts—the northern, however, is slightly larger.

The Milky Way, no less than other stellar phenomena, attracted the early attention of the people. The Dieri Tribe in Central Australia says that the Milky Way is the stream of heaven and the Mexicans consider it the source of all that is. Tradition endeavoured to explain its origin. Its milky appearance caused the Romans to call it "*Via Lactea*," a name that is retained in translated form in most modern languages. This name is coupled with the legend of the Hercules-child, who sucked the breast of Juno and when it was pushed away by the incensed goddess, the milk was spread across the sky.

Nevertheless, the human race, until about two hundred years ago, had little conception of the extraordinary importance of the Milky Way. Anaxagoras and Democritos surmised, however, that it consists of a

---

[1] The literal translation of the Swedish name is "The Wintry Way."

collection of exceedingly minute and densely clustered stars each of which has the nature of our Sun. Ptolemy described, nearly two thousand years ago, its position on the firmament and his observations are valid today as far as determinations with the naked eye suffice. When Galileo introduced the telescope, the conception of the Milky Way as made up of innumerable stars was verified. Not quite two hundred years ago Swedenborg, in his cosmological speculations, considered our solar system as a part of the Galaxy. Wright, Kant, and Lambert further amplified these theories.

The first important forward step was taken by the great William Herschel in his statistical researches. He demonstrated that the stars lie closer to each other the nearer the Milky Way they are located. This is mainly true about the small stars invisible to the naked eye while the more luminous ones are more evenly distributed over the heavens. In certain parts of the Milky Way the stars are over one hundred times more crowded than at its poles—that is, the points farthest removed from the Galaxy. Herschel's investigations were continued and elaborated by Struve, and later by numerous other scientists.

Through these researches, it has been determined that the Milky Way is, so to speak, the foundation upon which the star system, visible to us, is built. All kinds of stellar bodies have been studied and their distribution has proved to be symmetrical with reference to the plane of the Galaxy. The majority is greatly concentrated around the Milky Way. To these belong the new stars which occasionally blaze into existence, as the well-known new star in Perseus, 1901, and nearly all of which have appeared in the Milky Way or in its immediate vicinity. We also find there the irregular nebulæ, enormous, vastly diffused volumes of gas, among which the best known is the Orion nebula, and which seem to constitute the primeval matter out of which the stars are born. We might further mention the star-clusters, dense, ball-shaped agglomerations of stars, and the so-called planetary nebulæ, which—at least in their visible outer shell—also consist of gas accumulations with a spherical or ellipsoid conformation. The numerous spiral nebulæ on the other hand, those strange stellar bodies to which we later shall have occasion to return, are beyond comparison more frequent in the regions surrounding the poles of the Galaxy than in the rest of the heavens.

Many astronomers have considered the Milky Way itself a nebula. The most common theory doubtless is that it closely conforms to a spiral nebula—an opinion that has found a particularly warm advocate in the Dutch astronomer, Easton (see Figure 1). A few years ago Prof. Bohlin expressed the view that it is most akin to a planetary nebula, or more precisely to a ring nebula which is supposed to grow out of a planetary ellipsoid nebula by the gaseous matter being driven from its poles toward its equator. It is of a certain interest that this theory lends itself to the support of Swedenborg's—nevertheless improbable—hypothesis about the origin of the planets in the solar system. As we later shall see the Easton conception has the better reasons in its favour.

If classified according to age the stars are again distributed with the Milky Way as a46 reference point. Thus, let us consider their evolution, which for various reasons is assumed to take the following course. We may commence when the star-matter existed on the nebula stage. It then radiated the light characteristic of certain incandescent gases, principally the lightest two, hydrogen and helium, and further of an otherwise unknown gas called nebulium (nebula-substance). These gases were later condensed and dark spectral lines commenced to appear beside the bright lines of the aforesaid gases. Stars on this stage, named after their discoverer Wolf-Rayet stars, occur only in the immediate vicinity of the Milky Way. A later stage in their evolution is represented by the so-called helium stars in whose spectrum the dark helium lines preponderate. They are considerably concentrated around our Galaxy. Somewhat more evenly distributed and yet of decidedly greater frequence in the neighbourhood of the Milky Way, the hydrogen stars appear, characterized by strongly developed hydrogen lines and somewhat retreating helium lines. These stars are more developed than the helium stars and form with them the group of white stars so named after the colour of their light. Next in evolution follow the yellow stars, to which our Sun belongs. Dark metal lines appear in their spectrum. They are more evenly distributed than the groups mentioned before. Still further is this true about the red stars whose spectra contain the characteristic bands of chemical compounds and therefore betray comparatively advanced cooling. They are fairly uniformly spread over the heavens but are still somewhat more numerous in the vicinity of the Milky Way than further therefrom.

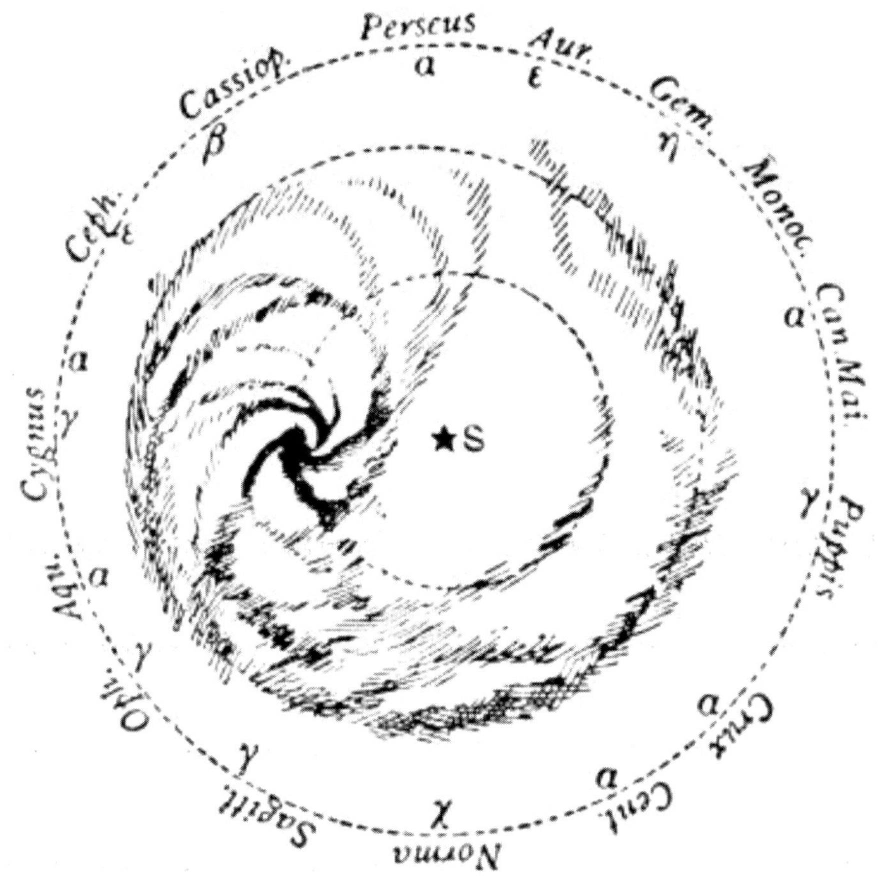

Figure 1. The Milky Way, pictured as a spiral nebula by Easton.

These facts are demonstrated in the statistics by E. C. Pickering, Director of the Harvard Observatory, who divided the firmament in four equal zones, the first of which is nearest to the Milky Way (and includes it) and the last of which contains the Galactic poles. His table shows the percentages of different stars in each of the four zones.

Galactic Latitude	Helium Stars	Hydrogen Stars	White Yellow Stars	Yellow Stars	Red Stars
±8.1°	51.2	37.4	29.7	29.4	26.7
±21.6	31.7	28.6	27.9	26.7	27.6
±39.8	11.9	18.3	21.1	21.9	23.6
±62.3	5.2	15.7	21.3	22.0	22.1
Number of stars observed	716	1885	1329	1719	457

The difference is most pronounced in the two first groups; in the three last it is small but unmistakable. An even distribution would correspond to 25 percent in all four divisions of the heavens.

These comprehensive statistics, embracing 6106 stars, seem to indicate that the stars in their first stage were within the Milky Way but subsequently drifted away with increasing age. This leads us to the thought that they originated from the irregular, nebulous accumulations which occur in the Milky Way and in its vicinity, or more correctly from similar formations which formerly existed in these regions but which now have clustered into stars. This agrees very well with another observation. With the help of the spectroscope the motion of different stars has been determined with reference to the point where the sun now is. The velocities have been found greater the older the stars are as shown in the table below taken mainly from the investigations of the renowned astronomer, Campbell.

Mean velocity of:				
Irregular nebulæ	0	Km.	(0 Miles)	per sec.
Wolf-Rayet stars	4.5	"	(2.8 ")	" "
Helium stars	6.5	"	(4.0 ")	" "
Hydrogen stars	11	"	(6.8 ")	" "
Yellow stars	15	"	(9.3 ")	" "
Red stars	17	"	(11.5 ")	" "
Planetary nebulæ	25	"	(15.5 ")	" "

To these figures a few remarks founded on recent observations might be made. The average distance between us and the stars in each group is different and the yellow stars, to which indeed our Sun belongs, are those nearest to us in space. They are therefore easier to observe than stars in the other groups. Campbell's statistics include also for this reason a smaller number of stars in this class than in the others. It is conceivable and by the astronomer Halm held to be true that the mean velocity of the smaller stars is greater than that of the larger ones. This is the condition existing in a mixture of different gas molecules, with which the brilliant French scientist, Henri Poincaré, compared the throng of stars, inasmuch as the heavier molecules possess the slower motion. To confirm this W. S. Adams of the Carnegie Observatory on Mount Wilson compared stars of equal velocity in their own orbits. Such stars are considered to be on the average equally far removed from us. He found the theory of Halm confirmed. The mean velocity of the hydrogen stars was reduced from 11 km. (6.8 miles) to 7.5 km. (4.7 miles), that of the yellow stars from 15 km. (9.3 miles) to 9.2 km. (5.8 miles), and that of the red stars from 17 km. (11.5 miles) to 14 km. (8.7 miles), while that of the helium stars remained unchanged. The sequence of the stars arranged according to velocity in the line of sight is evidently not modified by this new calculation.

In regard to the motion of the planetary nebulæ it should be mentioned that Campbell in this connection has carried out a great number of new determinations according to which the mean observed velocity of these large bodies in the line of sight must be increased to not less than 42 km. (26 miles) per second.

Campbell and Moore contribute the following interesting data in regard to Nebula N. G. C. 7009 (Figure 2):

"Measures of the rotational velocity of the nebula enable us to draw some interesting conclusions concerning its mass. On the most plausible assumption as to the location of the axis of rotation the orbital speed of the nebular materials lying at a distance of 9 seconds of arc from the centre is about 6 km. (3.7 miles) per second. If we provisionally assume the mass of the central nucleus to equal that of the Sun, Kepler's law connecting the periodic time with the distance from the nucleus tells us definitely that the nebula is distant from us only 8.9 light years. This must be regarded as an improbably small value, in view of other evidence bearing on the question. For assumed distances of 100 and 1000 light years, which we have reason to believe are more probable orders of nebular distance, the masses of the nebula would be respectively 11.3 and 113 times that of the Sun, and the corresponding periods of rotation 1371 and 13,710 years. From these considerations it seems certain that the mass of the planetary nebula N. G. C. 7009 is several times that of the Sun. The nebula is therefore competent, from the point of view of its mass, to develop into a system more pretentious than is our solar system.

"A few speculations concerning this nebula may not be without interest and value.

"The faint extensions to the east and to the west of the elliptical figure suggest an encircling ring of materials whose principal plane, passing through the nucleus, passes also near our (the observers') position in space. These extensions terminate in condensed nuclei at equal distances from the nucleus and on exactly opposite sides of the nucleus. The faint extensions and condensations may be and probably are largely the effect of the edgewise projection of such a ring, as in the case of Saturn's rings when the observer is in the plane of the rings. The forms of the two terminating condensations, and especially the wing extending up and out from the east condensation, suggest that we are not precisely in the plane of the assumed ring.

"The form of the main nebula appears to be ellipsoidal and not chiefly elliptical.

"The space immediately surrounding the central nucleus appears to be relatively vacuous. Aside from the nucleus, the principal mass of visible nebulosity exists in the brilliant ring, roughly elliptical as to its inner and outer boundaries, which occupies the region about midway between the nucleus and the outer edge of the nebular structure. The brilliant ring is probably in reality an ellipsoidal shell; the projection of such a shell upon a plane at right angles to the line of sight would naturally show a relatively dark central area, but the projection principle may not be the only one involved.

"If this nebula is in process of development into a solar system, the indications are for a system having certain resemblances to our solar system. Our four outer planets have a combined mass 225 times as great as that of the four inner planets. Similarly in N. G. C. 7009, there is apparently a paucity of material to form planets near the nucleus and an abundance of material for planets at greater distances from the nucleus."

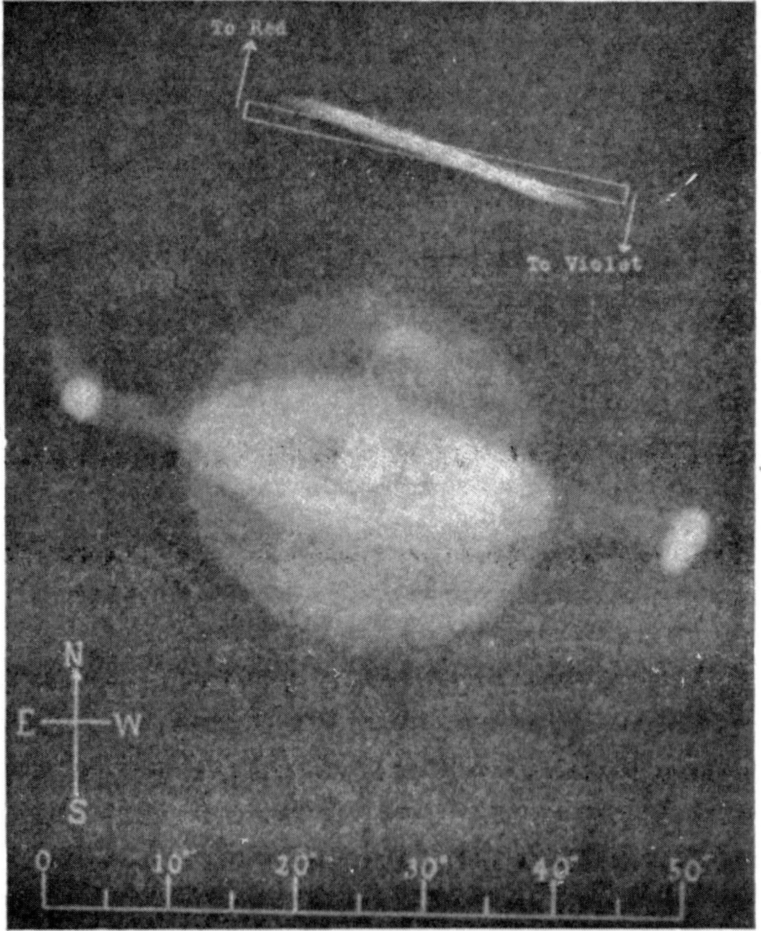

From Proceedings of the National Academy of Sciences of U. S. A.

Figure 2. Planetary nebula N.G.C. 7009 (composite drawing, from Curtis's photographs of the nebula made with the Crossley reflecting telescope. The scale is in seconds of arc).

Interesting observations have been presented recently also with reference to the largest among the irregular nebulæ, namely the Orion nebula. Three astronomers in Marseilles, Bourget, Fabry, and Buisson, found that parts of this nebula, in the neighbourhood of the so-called trapeze and very close to each other, moved with different velocities and that this difference might amount to 10 km. (6.2 miles) per second. The south-easterly part approaches us while the north-easterly recedes from us. Consequently a violent whirl-motion undoubtedly takes place in this region. This observation has been verified by the well-known Chicago astronomer, Frost, who employed a different method of investigation than his predecessors. He noted differences in velocity amounting to 11 km. (6.8 miles) per second between points not over two seconds of arc distant from the trapeze.

If therefore we say that the irregular nebulæ on the average possess no motion, this statement does not preclude important local deviations from the rule within the nebulæ, intimating a transformation which probably leads to the concentration of the nebulous matter toward the centre of the whirl.

Leaving out, to begin with, the planetary nebulæ, it appears that the original matter of the stars stands still in space, that their average velocity increases with increasing age and approaches a mean value of about 18 km. per second or roughly 1000 times the speed of the ordinary passenger train. Our Sun, in particular, moves with a velocity of 20 km. (12.4 miles) per second toward a point in the constellation Hercules 30 degrees north of the equator.

What force then shall we say it is that causes the motion of the stars? As far as we know none but gravitation. It appears therefore as if the gaseous primeval substance of the stars were not governed by this force. It might prove hazardous, however, to make this assumption as gases also possess weight and even the most rarefied strata of the Earth's atmosphere exert barometric pressure by virtue of their attraction to the mass of the earth. Rather the immobility of the nebulæ is due to the frequent collisions between the molecules in any quantity of gas even if it be attenuated to such a high degree as in the nebulæ. Thus, the molecules strike a balance, as it were, against each other so that the different parts of the gas accumulations shortly are brought to rest relative to each other. The irregular gas mists around the Milky Way form therefore a continuous whole. A different condition obtains in regard to condensed stellar bodies such as the stars. They may in the densest throng move during billions of years before they collide; but they might on the other hand enter nebulous masses and thereby suffer gradual retardation. We now refer to stars moving outside of the vapour clouds. They are therefore unrestricted and the longer they have obeyed gravitation without impeding encounters with nebulous matter, in other words the longer the time elapsed since they emerged from the gas accumulations which gave them birth, the swifter is their motion. Their (average) velocity can of course not exceed a certain limit which in our parts of the universe appears to be about 18 km. (11.2 miles) per second. Campbell's measurements show that for the youngest stars (all except the red) the velocity is greatest in the plane of the Milky Way, a natural enough condition as the attracting matter here is most abundant.

The planetary nebulæ possess a greater velocity although they, as consisting of mist vapours, are in the first stage of evolution. Faster yet do the spiral nebulæ move according to measurements by Wolf of Heidelberg. This shows that they are of a different nature from the irregular nebulæ, which form the matrix of the Milky Way. A closer examination of the few—thirteen in all—planetary nebulæ, determined by the American astronomer Keeler, convinced me that they approach the Galaxy from its poles with a moderate speed, and subsequently under the influence of its attraction curve their orbit, rapidly gain in velocity, and finally rush into the nearest part of the Milky Way with a very high speed.

A great number of them are no doubt caught in the mists or star-throngs of the Milky Way after exposure to numerous collisions and sweeping away all matter in their course. Such clean-swept traces are very common in the area of the Milky Way. One of the most beautiful examples is the so-called Cocoon nebula in the constellation Cygnus (the Swan). It has left in its wake a dark rift, in whose bottom, however, exceedingly small and evidently very remote stars are visible according to the German astronomer Wolf (see *Worlds in the Making*, page 172, Figure 55).

The great mean velocity of the planetary nebulæ indicates that they originally did not belong to the Galactic system, a conclusion also reached by Bohlin, but for other reasons. They are nevertheless more abundant in the neighbourhood of the Milky Way than in other parts of the heavens. This fact, if viewed superficially, might lead to the belief that they are indigenous to the Galactic system, but is explained by their concentration in obedience to gravitation toward the Milky Way.

Quite recently (1917) Van Maanen determined the distance of one of these highly interesting celestial bodies, tabulated in the New General Catalogue as No. 7662. Its distance was found to be only about 140 light years. This is about sixteen times the distance of Sirius and the mean distance of a star of the fifth magnitude. This circumstance agrees very well with the idea that this nebula is captured by the Galactic system to which it has approached from very distant parts of the space outside of the Galactic system.

One of the most remarkable astronomical discoveries in recent years was made by Kapteyn, who thereby as well as by other achievements has gained perhaps the highest rank among astronomers of today. He has shown that the stars rushing forth in the neighbourhood of the Sun belong to two great groups, one coming from the constellation Orion, and the other at nearly a right angle (100°) from the constellation Scorpio. In the former, we find nearly all the helium stars hitherto studied. We have previously seen that these stars stand almost still with reference to the Galaxy while the irregular nebulæ possess no motion at all relative to the same reference point—and the Galaxy is the natural datum-line for all astronomical measurements—so that the motion of the first-mentioned star-group toward the Sun is principally due to the Sun's own motion. This group, according to Kapteyn, obeys the law of relative star velocities even better than the combined world of all stars; thus with reference to the Sun, the helium stars are the slowest, the yellow stars the fastest, while the hydrogen stars occupy a middle position, all a self-evident consequence of their own velocity with reference to the Galaxy which increases from helium stars to yellow stars.

Kapteyn has shown another regularity in this group which is easily explained. We have previously mentioned that the yellow stars are most, the helium stars least removed from their place of birth in the Milky Way. The result is that the yellow stars appear (on the average) to come from a point farther from the Galaxy than the apparent origin of the hydrogen stars and more remote yet than that of the helium stars. On account of the relatively high velocity of the yellow stars in different directions, their stream appears to be more divergent than the stream of hydrogen stars, and helium stars move in almost parallel paths (nearly directly opposed to the Sun's true motion with reference to the Galaxy).

Similar regularities have been found by Kapteyn in the second star-drift which would lead us to think, as indeed Kapteyn assumes, that these stars also developed from an original nebulous mist, which arrived in our neighbourhood from the unknown distant, but is now used up in the formation of corresponding stars. Here again the yellow stars should have departed farther from their matrix than the white hydrogen stars. Helium stars are very rare in this drift, so that no reliable statistics have as yet been made for them.

It has been one of the most difficult problems of cosmogony to form a theory to account for the origin of the Galactic system. We may, almost yearly, witness how new stars blaze into existence only to fade rapidly and in a few years return to their old insignificance—that is, they become invisible to the naked eye although through powerful lenses we may frequently discover an exceedingly faint star in their position. As a rule, a nebula of the planetary type is formed in the course of a few months. Somewhat later the nebula is transformed into a Wolf-Rayet star. It is interesting to note that Wright found the central bodies in certain planetary nebulæ to be Wolf-Rayet stars. We have good reasons to assume this blazing forth into light to signify the collision of two faintly luminous or possibly extinct stars. The new lights appear also in stellar regions where the star density is very great, particularly in the Milky Way or its vicinity.

We see therefore repeatedly how mists with enclosed central stars originate. They remind us to a certain extent of the Galaxy with its clouds and stars and along the road thus suggested trials have been made to reach the solution of the riddle. The difficulty lies in the fact that the orbs whose collision create "new stars" are small, probably smaller than our Sun, while the mass gathered in the Milky Way most likely is trillions[2] of times greater than that of the Sun. It is true that we know a few unique stars, such as Arcturus, which exceed our Sun in size several tens of thousand times, but not even two such stars would account for the mass of the Galaxy, and furthermore the probability that two stellar bodies of such rare dimensions would collide is so very small that it must be left out of account.

Kapteyn's star-drifts, containing many thousands or probably millions of stars, appear to furnish the bridge that leads to the solution of the riddle of the Milky Way. These drifts were once enormous gas-clouds, in mass probably several million times greater than that of the Sun. They also had an extension equal to trillions of stars. The probability for the meeting of two such gas-drifts is comparatively large and should not be much smaller than for the entrance of a star-drift into the Milky Way, an occurrence which actually has happened as shown by Kapteyn.

When two such enormous gas-clouds meet, each with a cosmic velocity of about 20 km. per second, a long time would not elapse before the gas molecules in the region of interpenetration would be retarded in their original motions. An extraordinarily strong concentration and heating would occur in this territory, which is surrounded by the comparatively cold and heavy masses which remain unaffected because outside of the impact-area. A certain degree of equalization would naturally take place in the layers adjoining the boundary between active and inactive parts and the former would, furthermore, be set into a rapid spin around an axis perpendicular to the plane containing the two original motions. On account of the great viscosity of gases, particularly at high temperature, the central part would rotate as a coherent unit. Thus it would form a disk of gaseous

---

[2] American and French numeration; billions acc. to Swedish and English usage.

matter. This disk would be thickest in the middle and would become thinner toward the edge where centrifugal force acts most powerfully.

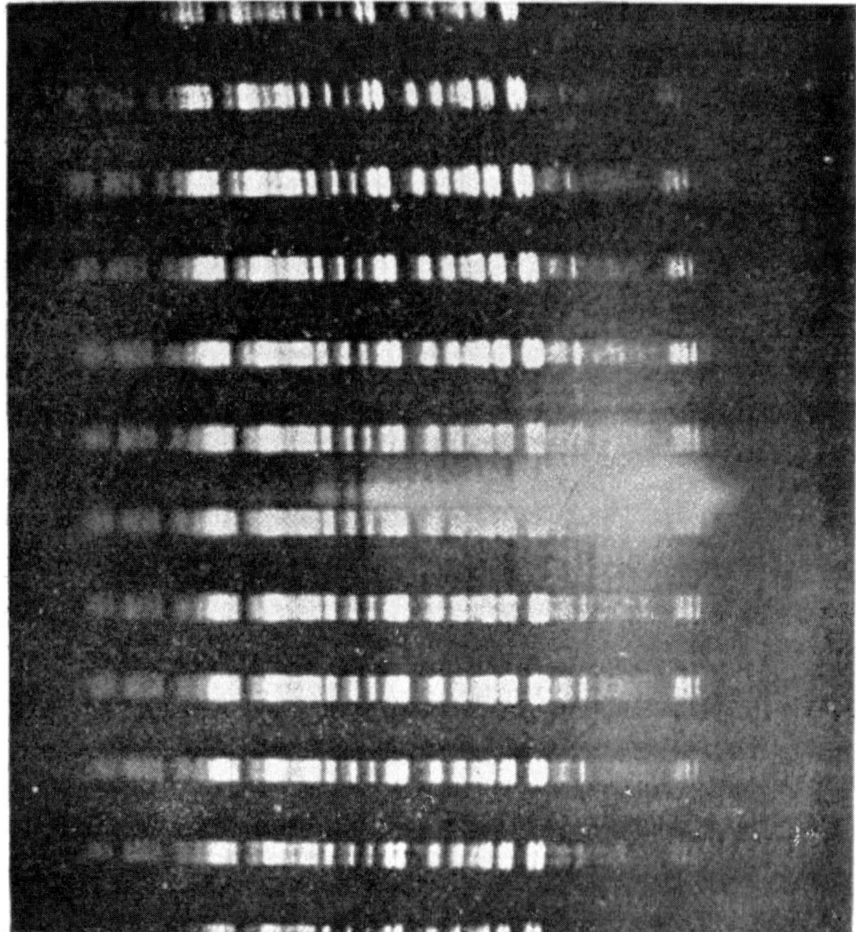

From Proceedings of the National Academy of Sciences of U. S. A.

Figure 3. N.G.C. 4594. Exposure = 2 hours. 1mm = 6”7.

Such a discous nebula has been investigated by the astronomer F. G. Pease of the Carnegie Observatory on Mount Wilson. By means of the spectroscope he has studied the motion of Nebula No. 4594 in the New General Catalogue (see Figure 3). This body is believed to be a spiral nebula like those in Figs. 4 and 5 but viewed from the side so that the spirals appear as a band. As the picture shows, this band is coursed through by a thick dark line, owing, it is believed, to a cold non-luminous dust-cloud outside of the spiral. The bright band is broadest in the centre. The curvature of the dark middle line in the shape of an arc, whose apex points downward, combined with the fact that the major portion of light falls above this arc, indicates that we do not view the nebula exactly on edge but from the upper (north) side of a plane through the arms of the nebula. The nucleus speeds away from us at the dizzying rate of 1180 km. (730 miles) per second. The east edge, *i. e.* the left on the picture, departs with the still higher velocity of 1630 km. (1100 miles) per second, while the west (right) edge retreats at the rate of only 800 km. (495 miles) per second. According to Pease, the nebula rotates as a solid disk so that the difference between the velocity of any point and that of the centre increases in the same proportion as the distance of the point in question from the centre. It is probable that we have been prevented from observing the outer parts, corresponding to the spiral arms proper, by the ring of dust which encircles the nebula. The visible portion occupies an arc of 2¼ minutes on each side of the centre. Its spectrum corresponds to that of star-group F-5 among the yellow stars in the Harvard classification. Therefore, it is not the light of the coherent gas-cloud which preponderates but rather that of the stars consolidated within the cloud, and corresponding to the stars in the Milky Way. This star-light is so bright that it entirely suppresses the radiation from the gas-cloud itself.

Such parts of the gas-aggregation as are most removed from the place of collision continue in their course through space little affected by the attraction of the central mass because of the great distance involved. The portions nearer the point of impact receive orbits curved by this same attraction and the curvature becomes the sharper the nearer the axis of rotation. One result of the mutual gravitation between the central mass and the particles in the outer sections of the nebula is also that the velocity in the spiral arms becomes smaller the farther the section in question is located from the centre, just as comets in the solar system move slower the farther they are removed from the Sun. But in all portions outside the central region the attraction is too weak to give circular orbits to the gaseous matter. All substance in these localities, therefore, departs ever more from the centre. As the spiral arms stretch out into straight lines such matter finally leaves the central disk altogether. It is possible that only the disk itself remains in the nebula computed by Pease.

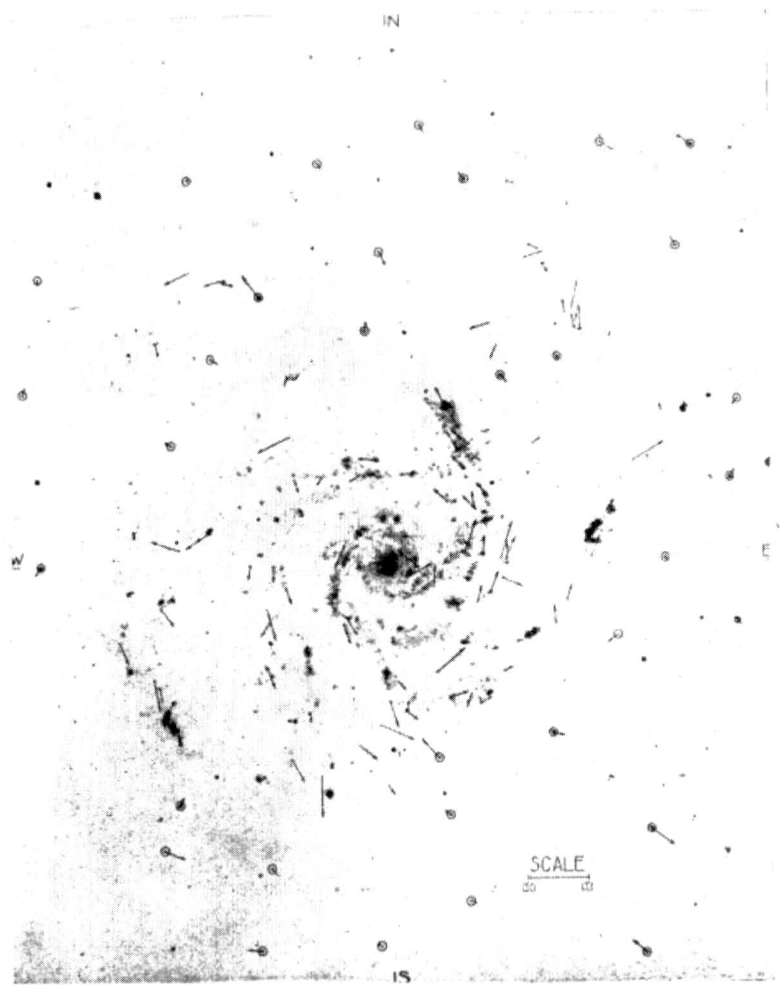

From Proceedings of the National Academy of Sciences of U. S. A.

Figure 4. Internal motions in Messier 101. The arrows indicate the direction and magnitude of the mean annual motions. Their scale (0"1) is indicated on the plate. The scale of the nebula is 1mm = 10"5. The comparison stars are enclosed in circles.

Another astronomer on Mount Wilson, A. Van Maanen, has investigated a nebula, No. 101 in Messier's catalogue (Figure 4). This nebula lies nearly at a right angle to the line of vision which consequently almost coincides with the axis of rotation. The motion of the different parts of this nebula has been calculated with the help of photographs taken in the years 1899, 1908, and 1914, whereby its changed position with reference to surrounding fixed stars has been recorded. Out of 87 points in the spirals only 9 moved in the direction of the hands on a clock while the other 78 moved in the opposite sense. The mean angular velocity is 0.022 seconds of arc per year, corresponding to 85,000 years for one complete revolution at 5 minutes' distance from the centre. The absolute velocity 2 minutes of arc from the centre is 1.5 times as great as at 7.5 minutes' distance.

Figure 4 is reproduced from Van Maanen's original. It shows clearly the general regularity in the motion of the component parts as well as the numerous exceptions to the rule. Such exceptions may be caused by perturbations due to invading masses, which impart their own motion to the entangling matter. These foreign bodies have probably condensed surrounding vapours and this created the bright knots which stud the nebular spirals. The upward motion amounts on the average to 0.007 seconds per year. While the points of condensation describe half a revolution around the centre they depart therefrom to about twice their original distance. More than a million years is therefore likely to elapse before the outer portions of the nebula are so far removed from the nucleus that the spiral form of the nebula is no more apparent.

It is clear that the Milky Way may have been formed through the collision of two immense nebulous gas-aggregations in the manner just described. Subsequently and by virtue of the magnitude of the Galaxy great quantities of wandering cosmic matter and minor stellar bodies have been gathered in occasionally accompanied by larger clusterings such as the planetary nebulæ referred to.

How well justified we are in looking upon the Milky Way as a spiral nebula is apparent from the picture (Figure 5) reproducing a photograph of the familiar regular nebula in the Dogs (Canes Venatici). It shows a wealth of detail hitherto not surmised. The feat was accomplished in the Carnegie

Observatory on Mount Wilson in Southern California with the help of optical resources vastly superior to all earlier means. The Milky Way has previously been compared to this nebula, but due to deficient enlargement their striking similarity has not been fully recognized until now.

Assume the Sun located in the point marked "S" in Figure 5 and some distance above the plane of the picture, then the nebula if viewed from this point would appear somewhat as the Milky Way appears to us. In the middle we behold the substantial nucleus and on its left side a cleft between the two branches of the inner spiral. Farther to the left, we see only the outer spiral, first broadening toward the left where it approaches "S," then narrowing only to spread again on account of the great clustering in the lower right part of the spiral. The axis of the nebula corresponds to the densest parts of the Galaxy in Cygnus (the Swan), the loop in the inner spiral again to the empty space between Cepheus and Cassiopeia, the narrow part of the outer spiral branch resembles the constriction at Algenib, the following diffusion corresponds to the broad section in Auriga (Charioteer) and Monoceros (Unicorn). At the subsequent narrow place we see the outer nebula-clump corresponding in certain respects to the Magellanic clouds on our firmament although these are farther removed, and apparently not indigenous to the Milky Way. There follows in the nebula a massive section in our system represented by the well-known, far less compact, yet brightly luminous tract containing the Southern Cross. Here, from the star Alpha in Centaurus—the nearest bright fixed star[3] to

---

[3] Up to a short time ago Alpha Centauri was considered the fixed star nearest to the Sun. By comparing old photographs of the firmament with such of recent date the renowned astronomer Barnard found that a very small fixed star—of the magnitude 10.5 and therefore far from visible to the naked eye—in the constellation Ophiuchus (Right ascension 17 h. 58 m. 44 s., North declination 4° 27′.4 January 1, 1917) possesses a very large proper motion. It traverses in a year 10.3 seconds on the firmament. The distance to this star, which has the largest proper motion so far known, was later determined to be 3.3 light years or 3/4 of the distance from Alpha Centauri to our Sun. Hence its velocity at right angle to the line of vision is computed to be 49 km. (32 miles) per second. Spectroscopic measurements show that it approaches us with a velocity of 91 km. (56.5 miles) per second along the line of vision. The combined velocity, therefore, is 103 km. (63 miles) per second, an unusually high value. The value of 3.3 light years used in this calculation was determined by the French astronomer Gonnessiat, who found it by the study of old photographs from Algeria. He also calculated the parallax of this star to be 1 second. According to later measurements,

Figure 5. Spiral nebula No. 51 in Messier's Catalogue; situated in Canes Venatici, and photographed February 7 and 8, 1910, from Mount Wilson Observatory in California. Scale: 1 millimeter = 5 sec. arc.

---

given in the Harvard Bulletins 616 and 617, its parallax is only 0.7 seconds and consequently its distance 4.6 light years and its speed perpendicularly to the line of vision 70 km. (43.5 miles) per second. Campbell, in the Lick Observatory, had determined its radial velocity and found that it approaches the Sun at the rate of 128 km. (79 miles) per second. Its total velocity is, according to these two last determinations, 146 km. (91 miles) per second.

It is by no means improbable that similar discoveries will be made in the future, so that the Sun will be found to have more stars in its "immediate" vicinity than previously assumed.

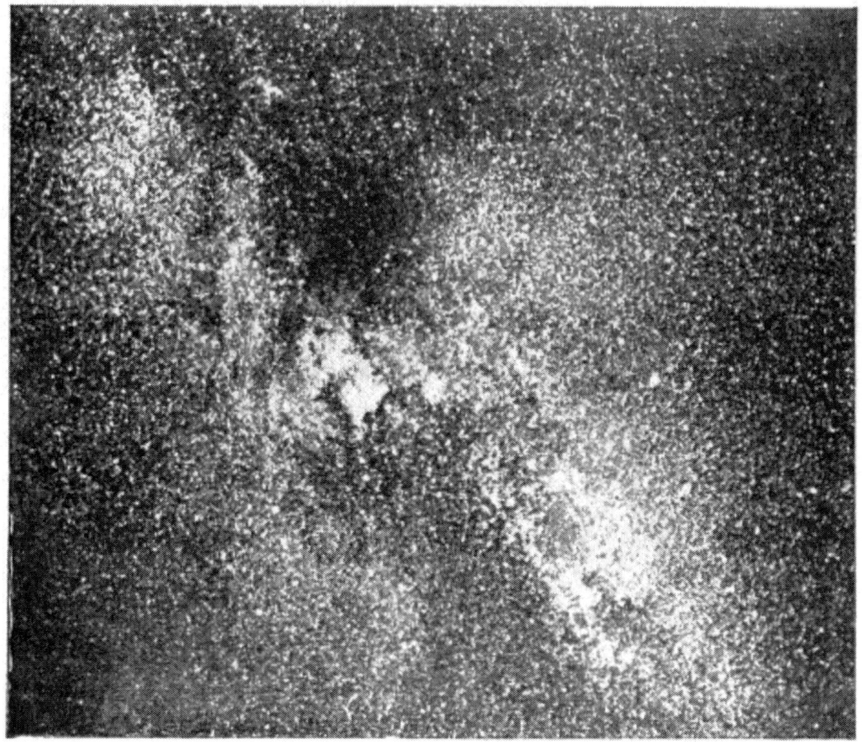

Figure 6. The Milky Way between constellations Cassiopeia and Swan from photo by M. Wolf of Heidelberg. A little to the left of the middle the beautiful nebula America appears.

our Sun, "only" 4.5 light years or about 25 trillion[4] miles distant—commences a bi-section of the spiral and strangely enough the nebula is similarly forked. Now the outer spiral stretching in a faint line upward from the "clump" begins to show as a weak band, while the inner spiral stands forth powerfully above "S" corresponding to the brilliant section of the Galaxy in Scutum (Shield) and Aquila (Eagle). The partition in the nebula between these two branches is the counterpart of the 110° long "prong" in the Milky Way between the constellations Norma and Lyra. Numerous faint bridges join the two branches in the nebula as well as in the Galaxy, according to Wolf.

[4] American and French numeration; the Swedish and English usage is billion.

The correspondence is indeed surprisingly good. Proportions are, of course, somewhat different—in particular is the central part of the Galaxy not so predominating, which fact has been troublesome to the adherents of the nebula theory. Probably it was originally denser but has become attenuated through star-formation, explaining, for instance, the great gap between the constellations Lyra and Vulpecula.

To give a better idea of the structure of the Milky Way, two photographs are here reproduced as taken by Wolf, the German astronomer in Heidelberg, who has done particularly meritorious work in this department. One (Figure 6) shows a section of the Galaxy in Cygnus (the Swan) with the star Deneb in the centre and to the left the "Northamerican-nebula" so named from its shape. Above Deneb is the dark "hole" in Cygnus and below another chasm not quite76 so black. Left of the "hole" is the winding canal enclosing the so-called Cocoon nebula. (See *Worlds in the Making*, page 172.)

The following picture (Figure 7) contains, in the upper left, the bright star Altair in Aquila (the Eagle) located close to the powerful arm of the Galaxy in this constellation. Farther to the right is the fainter arm in Ophiuchus (the Serpent-holder). The lower half contains the most brilliant part of the Milky Way in the constellations Scutum (the Shield) and Sagittarius (the Archer). Bright stars are infrequent but the fainter ones are innumerable: "They are crowded into dense clusters and between them the most delicate star dust is scattered."—"We behold how the star-ribbon dissolves into detached tufts which intertwine into the strangest patterns. These clouds of stars reach their greatest splendour in the lower part of the map."

We also reproduce (Figure 8) from M. Wolf a photograph in larger scale from the region of Gamma (lower part of Figure 7) in the Eagle with its "trident-hole," so called from its peculiar shape, and in whose vicinity mists and star-clouds abound. This picture is a more complicated

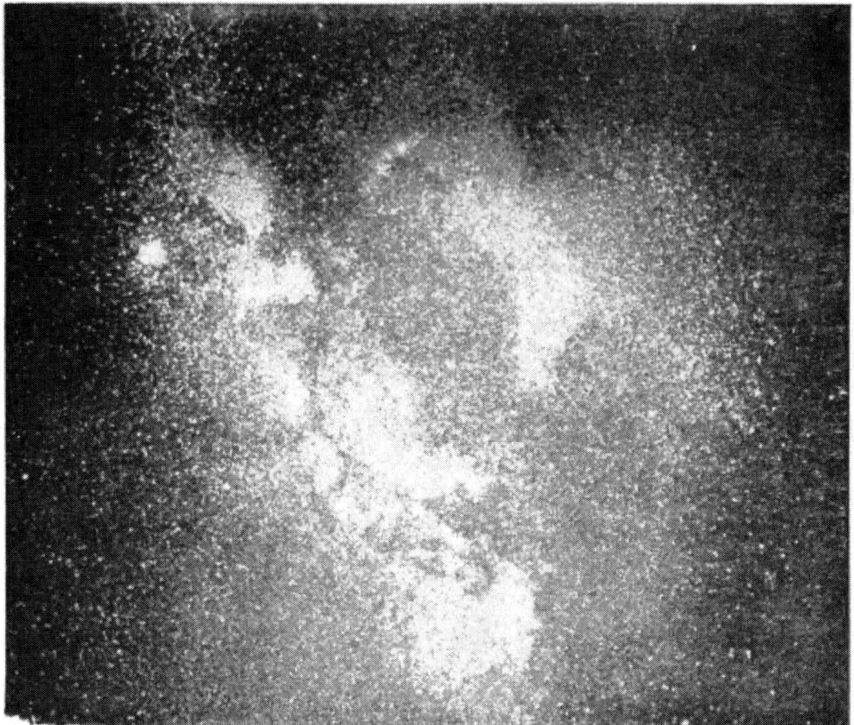

Figure 7. The Milky Way in constellations Eagle (upper half) and Archer (lower half). In the upper left the bright star Altair appears. Photo from M. Wolf of Heidelberg.

counterpart of the flatter photograph by Wolf of the "Cocoon nebula." It appears as if three or four stellar bodies here had stepped in from without, swept away the stars in their way, and left clean "streets" behind. Probably other "empty" spots in the neighbourhood have been formed in a similar manner. Another theory is that such dark places are caused by opaque mist-formations which shield the light of the stars behind from our sight.

Through these pictures we gain a conception of the manner in which the present stars in the Milky Way have clustered out of the original misty chaos. We cannot avoid the idea of great exterior similarity between the lumps formed in curdling or souring milk and those which we observe in the Milky Way. The renowned French scientist Duclaux says in his microbiology: "In milk, commencing to sour, but yet entirely liquid, we observe

Figure 8. The trisected hole (Trifid-hole) in Eagle. Photo from M. Wolf of Heidelberg.

under the microscope a precipitation of tiny particles. To begin with they are seen with difficulty and are discovered only by slightly displacing the plane of vision. Later they develop into distinct grains, characterized by Brownian movements, just like small particles of clay.... Still later the phenomenon appears as a steady molecular agglomeration. The grains have the tendency of the clay particles to lump and precipitate."

The first condensation-nuclei in the mist-clouds are no doubt cosmic dust entering from without and perhaps also larger clusterings such as meteorites and comets. At the existing low temperature, surrounding gases condense into fluid state on the dust particles which by virtue of these moist shells are cemented into aggregations of such size that gravitation overcomes the repelling radiation pressure. Gravitation assisted by the retarding vapours further mass these aggregates together. This process of coalescence is accompanied by heat production. Finally, small stars are formed, then groups of such stars, while the spaces between, now comparatively devoid of matter, appear dark much as the whey between flocks of curd. As yet, the small stellar bodies are surrounded by quantities of dust and gas, which, however, with continued condensation become ever more rarefied. Even yet the big stars in Pleiades, belonging to the helium group, appear on the photographic plates interspaced with great patches of dust-clouds. These are now, however, so unsubstantial that they offer little impediment to the procession of the mighty stars. The condensation process may be greatly accelerated through the invasion of voluminous gas nebulæ similar to the Cocoon nebula. At last all gases in the new star condense, that is to say the shell of tenuous vapours and dust contracts to such an insignificant thickness that it cannot be seen except possibly from the immediate vicinity. Small bodies ingathered through friction against the remnants of the original extended wrapping wander as planets around the new sun, sweeping away the last traces of unattached matter. The condensation on the new orb leaves a "hole" in the nebulosity which in this way is transformed into stars and their satellites which emerge from the mist and scatter on the firmament.

The Milky Way appears to be in a rather advanced stage of this evolution.

The "infinitely small" presents occasionally surprising likeness to the "infinitely large."

In this manner we can form a conception of the growth of the wonderful structure which has brought forth the majority of the stellar bodies that we discern. The spiral nebulæ visible at the Galactic poles are similar formations but probably of far more modest dimensions. They may compare to the Milky Way as the smaller planets to the Sun. According to recent investigations the spiral nebulæ seem also to possess an enormous velocity and they have probably invaded the Milky Way from without.

As previously stated, an exceedingly remarkable conception of the Milky Way exists among the Mexicans. To them it is the Matrix of all and gave birth to the stars, the most important of which are the Sun, the Moon, and Venus. This idea evidently agrees very well with the results of investigations in the last few years.

Finally, a few words about the extent of the Milky Way. So far we have not been able to measure it; only rather uncertain approximations are possible. Wolf estimates the diameter, that is the distance between the two spirals at the point where the Sun now is, to be about 10,000 times the distance from the Sun to the nearest fixed star, Alpha in Centaurus, which distance on the other hand is about 10,000 times the distance from the Sun to the remotest known planet, Neptune, or 300,000 times that from Sun to Earth. Expressed in the usual units, we arrive at 40,000 light years or 400,000 trillion kilometres (240,000 trillion miles). Lord Kelvin makes another estimate of 6000 light years, that is seven times smaller. The mean diameter of the nebula proper might be about five times larger, in round figures one hundred thousand light years, or one billion billion kilometres (600 million billion miles).

Like a monstrous octopus, the Milky Way swims in the fathomless ether-sea. Its dimensions are about as many times larger than those of the earth as that globe is larger than an atom. This has caused the gifted Irish physicist Fournier D'Albe to consider the celestial globes as atoms, out of which systems of the order of the Milky Way are constructed in the same manner as the earth and other stellar bodies are upbuilt with atoms, invisible to us and yet measured with an incredible degree of accuracy.

Fournier D'Albe goes further still. In poetic flight he does not hesitate to endow the Milky Way organism with life. Its evolution cannot be denied a certain similarity to the processes of life. The great nebula owes its origin to the union of two individuals, two nebulosities, who met in their course through boundless space. There floated the newborn extending its tentacles in the cool ether-waves and gained substance and strength through the smaller beings which the surging billows brought within its reach. It has now attained the zenith of its evolution, and is breaking up into molecules, or solar systems, which again are composed of stellar bodies, or atoms within the molecule. In violent exuberance of youth these rush through space in fulfilment of their individual life. Many will in due time undoubtedly become dust again and then serve to nourish a new youthful nebula. Others succumb to a freezing-death but will be restored to life in collision with a nebulosity or some other stellar body and give form to "new stars" or planetary clouds. Again and again shall the starry mists go through the cycle of existence and after a lifetime, whose duration stands in proportion to their dimensions, i.e., may be estimated to billions of billions of years, give rise to new celestial beings. Thus shall it forever continue in an eternal rhythm.

Chapter 3

The Climatic Importance of Water Vapour

When Aristotle, for two thousand years our leading savant in cosmography, about twenty-three centuries ago stated the foundation of his natural science, he laid down as the important principles: moisture and heat and their opposites; because the four elements out of which everything was made, were: earth, characterized by dryness and cold; water, which was moist and cold; air, which combined moisture with heat; and lastly fire, which stood for dryness and heat. Undoubtedly he was considering the requisites of life, which may be designated as humidity and heat. We have seemingly agreed that all life originates in the sea, so that moisture can be considered the first requirement for its appearance on the earth. As to heat, life is destroyed by frost and favoured by increased warmth, at least to a certain point, about 35° to 40° C. (95° to 104° Fahrenheit), which temperature is most propitious to the development of life, while a further increase is detrimental, so that already below the boiling point of water life suffers more harm than at temperatures below freezing. In fact, the geologists have found that the different epochs in earth's evolution are best characterized by their humidity or dryness. To arrive at a clear conception in these matters we shall briefly survey our present knowledge of the importance for the evolution of life on earth that we should attach to the humid and to the dry periods or localities.

We are all familiar with the heavy, moisture-laden warmth which meets us when we enter a hothouse. It is particularly favourable to the growth of plants and to the prosperity of the lower animals. To the higher animals and to man, the humid heat is not so beneficial. In the open such hothouse air exists only in the tropics. Particularly the Congo region and the parts of Brazil adjoining the Amazon River are remarkable for their humid heat and for their fabulously luxuriant vegetation. From our greatest living climatologist, Julius Hann, I have borrowed the following description of such a clime:

"The changes of temperature between the coldest and the hottest month are very small in the Congo, from .5° to 5° C. (.9° to 9° Fahrenheit), with an average of about 3.5° C. (6.3° F.). The difference between day and night reaches nearly thrice this value, or 9.5° C. (17.1° F.). The dry season becomes shorter the more we approach the equator and in Equatorville and Bangala it shrinks to nothing. During the rainless months, a dense humid fog settles morning and evening over the savannahs. Low hanging clouds of uniform thickness frequently hide the Sun for weeks at a time. It is during the rainy season alone that we see a clear sky between the showers. This season opens and closes with magnificent thunderstorms coming from the east. In Luluaburg, lightning occurs during not less than 106 days in the year. In the dry season the wind carries with it clouds of dust which falls to the ground. The cloudiness is enormous in the Congo basin, so that there are veritably no months with a clear sky in this part of the world. In Vivi, the number of overcast days averages 74 percent, fluctuating between 63 percent in August and 83 percent in November. The humidity is very high, varying in Vivi from 70 to 79 percent, with a mean value of percent, and in Bolebo the mean itself reaches 79 percent During the rainy season, the heat is sometimes unbearably oppressive; suffocating fumes rise from vegetable matter which rapidly decays in the excessive humidity. The annual precipitation does not reach very startling figures; it varies between 120 and 180 cm. (47 to 71 inches). In Gabun, close by, the sky is almost continuously covered with clouds during the dry season.

"Corresponding regions in South America are in parts characterized by an even higher humidity. In Iquitos by the Amazon River, it reaches not less than 83 percent of saturation. The annual change of temperature is only about 5° C. (9° F.); in Para (1.08° south latitude on the coast) it shrinks to 1° or 1.5° C. (1.8° to 2.7° F.). In the course of twenty-four hours the variation is considerably larger. The sky is remarkably clear between showers during the rainy season. In the interior of Guiana, the rains continue from the end of April well into July or even into August. Abundant dew is common during the rainless part of the year, thus maintaining the humidity. Sun and Moon are rarely visible, and gigantic lightning storms announce the arrival of the rainy period."

Similar conditions apparently prevailed during the carboniferous period, which was characterized by a luxuriant vegetation. The mighty tree-trunks of that time fell into the water-covered marshes out of which they had grown and their decay was thereby prevented. Instead they turned into coal like the peat in the mosses of today. This was for some time thought to indicate that the temperature was not particularly high—Frech estimated about 12° C. (53.6° F.) (1910). But since the discovery and subsequent description by Keilhack (1914) of peat-beds on Ceylon, where the average yearly temperature is 26° C. (78.8° F.), a return is to be expected to the older conception, which held that the vegetation during the carboniferous period is evidence of a very warm climate. Judging by the appearance of fossil plants, the temperature should have been nearly the same all over the globe. Carthaus remarks that the air was stirred by only feeble winds because the trees of that age with their great dimensions but frail root-systems could not have withstood a fresh breeze. The sky was hidden behind a continuous thick cover of clouds which only let a faint light sift through to the ground. The motionless air was almost saturated with moisture. The luxuriance of the vegetation, transcending anything existing today, indicates a favourable high percentage of carbonic acid in the air. This combined with the humidity and the dense clouds caused the heat radiation from the Sun to be almost entirely absorbed by the upper strata of the atmosphere in which thereby a strong circulation was maintained. As a result, the heat was nearly equalized between the poles and the equator and under the cloud cover an almost constant temperature reigned day and night, summer and winter. The damp air stood well-nigh still and was filled with dense fog at the smallest changes in temperature. Lack of light prevented the development of flowers, and the thriving plants belonged mainly to the ferns and to the horsetails. Pine and fir trees were yet comparatively few. The conditions in the swampy regions where plant life flourished were nearly identical with those in a hothouse if we were to draw a dense veil in front of the windows in walls and ceiling so that a continuous twilight would prevail.

In this uniform climate, plant life developed enormously faster than animal life. The dense clouds could store considerable quantities of heat in the equatorial belt through evaporation in their upper layers and the violent wind storms above the clouds would carry the aqueous vapours to colder regions where the heat would be liberated through new cloud formations. Currents in the oceans now largely attend to this heat transportation and give for instance to the coast of Norway, and indeed to the whole of Western Europe, its remarkably mild, and to life and civilization, propitious climate, but in the carboniferous age humid air currents fulfilled the same task. They moved considerably faster and more evenly than the ocean currents, were not checked or deflected by coasts or islands, and could therefore produce the extraordinarily uniform temperature and the marine climate all over the globe. Such a heat distribution takes place also in our days at a height of about 10,000 m. (6.2 miles) in the so-called "stratosphere," but the temperature here is very low, about -60° C. (- F.), so that the vapour suspended is hardly worth mentioning, and cannot give rise to cloud formations. The quantities of heat carried in these higher strata of the atmosphere are too insignificant to influence the masses of air below, whose temperature, therefore, is almost entirely governed by that of the sun-heated surface of the earth, except where the ocean currents equalize matters, as for instance in the almost wholly water-covered latitudes south of the 30th parallel on the Southern Hemisphere. Even during the carboniferous period at its height, there existed, of course, a temperature difference between pole and equator, but it was very small, some 10° C. (18° F.) perhaps. Undoubtedly, the formation of coal beds was mainly confined to those regions where the climate was most uniform all the year around.

The opposite extremity, the dry desert climate, is far more pronounced in the present time. This condition is well known in all continents except Europe, where we hardly can claim a desert but instead have steppes, with a vegetation abundant after the spring rains but fast disappearing with the arrival of the burning summer heat. A particular type of plant life has adapted itself to this periodic change from rain to drought, from bitter cold during the winter to parching sun during the summer. Perennial plants, and particularly trees, can rarely endure the rigors of such climatic upheavals. Animal life on the other hand has proved fairly adaptable and displays considerable wealth.

This steppe climate is only an intermediate stage towards the desert climate proper, which is hostile to all life. Its temperature is subject to enormous changes in the course of the day and the year. The annual variation is less pronounced near the equator and the daily variation less on the approach to the poles, on account of the small changes in the sun's radiation during corresponding periods. The difference between day and night in Sahara is frequently 30° to 40° C. (50° to 70° F.). The lowest temperature observed by Foureau-Lamy, 1898–1899, was -20° C. (-4° F.) or nearly the same as on the Scandinavian coasts. The highest amounted to 48° C. (118.4° F.) or a total variation of nearly 70° C. (126° F.). In Upper Egypt (21.9° N. Lat.) the mean temperature changed from 16.3° C. (61.3° F.) in January to 34.1° C. (93.2° F.) in July, and nearer the equator in Central Africa (8.1° N. Lat., 23.6° E. Long.) the difference amounted to only 6.9° C. (44.6° F.), 22.7° C. (72.5° F.) in December, 29.6° C. (85.1° F.) in April, while in Kiachta (50.4° N. Lat., 106.5° E. Long.) in Siberia, the yearly change reaches 45° C. (81° F.), -26.6° C. (-15.7° F.) in January, 19.1° C. (66.2° F.) in July. The average daily variation at continental stations is about 12° C. (21.6° F.). All this refers to the temperature of the air, while the surface temperature in the course of twenty-four hours may change 50° C. (90° F.) and in the desert even more. Frost occurs in the Sahara as late as May when the maximum temperature may reach 50° C. (122° F.). While in Scandinavia the diurnal difference between highest and lowest temperature averages only 6° to 7° C. (11° to 13° F.), a maximum in July of 10.4° C. (18.7° F.) and a minimum in November of 4° C. (7.2° F.), Hedin on his journey in Tibet, 1899–1902, observed a daily variation of 19° C. (34.2° F.) and no appreciable difference with change in altitude.

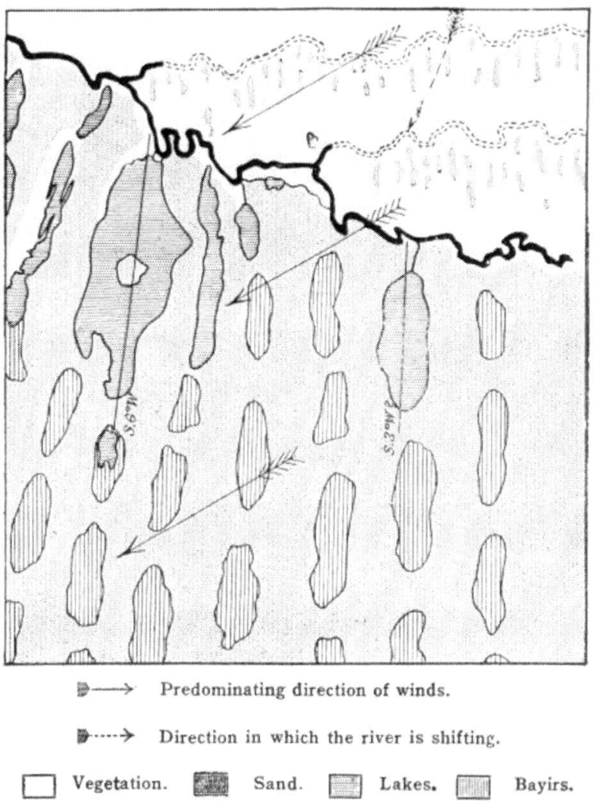

Figure 9. Tarim river with adjacent lakes and Bayirs, drawn by Sven Hedin.

(solid arrow) Predominating direction of winds.
(dashed arrow) Direction in which the river is shifting.
(hollow box) Vegetation.
(dotted box) Sand.
(horizontal lines in box) Lakes.
(vertical lines in box) Bayirs.

The result of such a violent temperature change in the course of a day is a breaking up of the rocks which subsequently and gradually are ground to fine dust by unobstructed winds wherever vegetation does not bind the soil. In this manner the sand deserts are formed. The arid wastes of Asia have lately been vividly described by Sven Hedin. The mountains eroded by the sandstorms resemble dilapidated ruins, standing as monuments of an ancient highland. The sand in East Turkestan is reduced to such a fine powder that it can float in the air for several days after a storm, revealing itself in gorgeous sunsets. Winds sweep the sand into long dunes, which shift in the direction of the blast. It is ferruginous and therefore red or if powderized reddish-yellow. When moistened it assumes a brown to black shade. After rain, the water descends toward the valley, carrying with it the sand in the form of silt. This, through evaporation, is transformed into a plastic black dough, slides like a glacier slowly down the hillsides, and finally comes to rest in some broad hollow which it gradually fills. Such a silt aggregation is called in Persia a "Kevir." Its surface dries, but the interior remains moist. As evaporation continues it becomes richer in salt so that white crusts of this substance are formed during dry periods. In other districts, as in the basin of the Tarim River, the water occasionally appears in the lowest parts, the so-called "Bayirs" (see Figure 9), formations similar to the Kevirs, or in salt lakes between the sand dunes. Sand carried by the winds quickly fills these lakes so that they too move in the direction of prevailing winds. They lie with their longest dimensions parallel to each other and at right angle to the course of the Tarim River. The sketch map taken from Hedin's work shows the Bayirs strung out in line with the lakes somewhat like panels in a tapestry pattern. This reticulation of the landscape is the result of the dune formations. The main dunes with steep western slopes run in the direction N.N.E.-S.S.W. They stand at right angle to the prevailing winds. Nearly perpendicular to their crest lines dunes of smaller height are thrown up by winds in another common direction but less frequent than those which raised the fundamental dunes. This system brings to mind the cloud formations called mackerel sky, clouds rippled in two directions frequently almost at right angle to each other. They owe their peculiarity to two series of wave motions propelled by winds from two different directions in the upper strata of the atmosphere. The cloud patches correspond to the wavecrests on a surging sea. The map of the Bayirs suggests a chessboard with squares somewhat elongated and irregular.

We may now return to a closer study of the largest of these formations, the great Kevir in Persia. This mud-lake, with a dry surface, measures 500 km. (310 miles) in length and 200 km. (124 miles) in width over its largest dimensions. Hedin estimates its area at 55,000 sq. km. (21,142 sq. miles) or the same as that of the great Lake Michigan on the boundary between the United States and Canada. Due to the continuous growth of the salt proportion through the inflow of Kevirs and through superficial evaporation a salt crust of varying thickness is formed near the surface of the lake. Hedin caused a hole to be cut with an iron bar. He first encountered a 10 cm. (3.9 in.) deep covering of clayey paste, and then the salt crust about 7 cm. (2.8 in.) thick resting on a semi-dry layer of clay with a depth of 15 cm. (5.9 in.). Farther below, softer strata of clay followed, becoming more watery the deeper he went. The iron bar carelessly wielded would have disappeared in the mire. Another investigator, Buhse, examined a piece of the crust, which when dry is fairly solid and of a yellowish-grey colour. One half consisted of sand (probably quartz-sand), one sixth of limestone, 6.1 percent oxide of iron (causing the yellow colour), 5.3 percent common salt, 2.5 percent sulphate of sodium, and 2.1 percent clay. Rain converts this surface layer into a plastic mass, which persistently sticks to the clothes of the traveller or to the bodies of the camels if they should slip and fall into the mud. Not the slightest trace of vegetation or of any life exists. On the shore of the mud-lake small flat elevations and depressions may be found; otherwise, the surface is as level as that of an ordinary lake.

The Kevir battles with the drifting sand as does the water in East Turkestan. The sand appears to gain in the contest. After storms, vast portions of the Kevir are covered with yellow desert sand. "If the climatic change in Persia continues in the present direction," says Hedin, who, however, is dealing with large spans of time inasmuch as in his opinion no appreciable alterations have taken place since the invasions of Alexander the Great, "then it may be taken for granted that the slough of the Kevir will lose in moisture and afflux of water and in time will become more solid and that the drift sand with greater ease will gain foothold and territory. The final outcome of the physico-geographical transformation now in progress will no doubt be to convert the entire Kevir into a sand desert of the kind predominating in East Turkestan. And conversely we may infer that East Turkestan, once a part of the central Asiatic Mediterranean sea, in the course of time was filled with the finely divided products of disintegration, such as we now find in the Kevir, and further that its expanse of watery mud and clay finally dried and hardened to such an extent that it could support the load of the encroaching sand. That the extension of the sand formerly was smaller, is also borne out by the archæological discoveries in East Turkestan which several travellers besides myself have made. The hardpan laid bare in the 'Bayirs' of the Cherchen desert strongly reminds of the Kevir-ground. In both cases the same dark, fine powder forming a nearly plane surface. In both cases this material when mixed with water is transformed into a slough in which one hopelessly sinks, but in East Turkestan the water has receded to a greater depth and as rains are extremely rare travel all over the smooth 'Bayir'-ground may be undertaken with impunity."

These formations are of the greatest interest because they picture the changes taking place on a slowly desiccating planet. In 1858 the Geographical Society in Petrograd despatched an expedition under the command of Khanikoff to visit these regions. From Hedin's work, *Overland to India*, from which the preceding quotation is taken, we borrow the following vivid description by Khanikoff: "At last, in the morning of April fourth, and during the most oppressive heat, we reached Bala-haus. At this place remnants of a ruined reservoir, long since bereft of its water, could be seen. The desert had here assumed the perfect character of the 'the accursed land,' which name it bears among the natives. Not the smallest tuft of grass, not a sign of animal life gladdened the eye, not a sound interrupted the deathlike, awful silence but that of the marching caravan.

"Because of the slow procession of the camels and the delay suffered when we lost our way, we only covered 25 km. (15.5 miles) in the night stage. After four hours' rest we resumed our march and directed our steps toward the hills, called Kellehper and situated 20 km. (12.5 miles) from Bala-haus; they were plainly visible but positively seemed to take flight on our approach. I was somewhat ahead of the caravan and sat down at the foot of this sand-elevation; and never can I describe the feeling of weariness and of depression that I was unable to ward off as I looked into the ghastly solitude that engulfed me. Scattered clouds intercepted the rays of the sun, but the air was hot and heavy; the diffused light lent a monotonous and disconsolating hue to the greyish, burning surface of the desert; hardly a single variation of colour gave relief to the immense expanses that the vision embraced. The absolute immobility of every point in this mournful landscape, combined with the complete absence of any sound, produced an overwhelming depression of the spirit; one felt as set upon a place that had been struck inanimate forever, a place to which organic life would never return safe through some terrible catastrophe of nature. One witnessed the beginning, so to speak, of the death-agony of our planet."

If a drying out has taken place in these regions,—which seems probable from Hedin's observation that the water in a Tibet lake, Lakker-tso, formerly reached a 133 m. (435 ft.) higher level than at present,—such a process is nevertheless not so obvious here as in the salt inland lakes, for instance the Great Salt Lake in Utah, the Dead Sea, and the Caspian Sea, where the saltness has greatly increased due to evaporation. Concerning the Great Salt Lake we know that even at a comparatively late time it had a much wider extension than now. Its water contains 22 percent common salt besides other compounds. The Dead Sea holds 25 percent salt. A very variable percentage is to be found in the Caspian Sea. Near the mouth of the Volga it is of course low, only 0.15, and increases southward to 1.32 at the peninsula Apsheron and to 5.63 in the bay of Kaidak. In the gulf of Karaboghaz on the Asiatic side it reaches 28.5 percent It has been computed that this gulf receives annually from the inrushing waters of the Caspian Sea 350,000 tons of salt which is partly deposited on its shores and bottom.

This desiccation, however, is a mere trifle when contrasted with the process by which the mighty salt deposits in Germany were formed. It took place, we believe, in a shallow bay extending southward from the Arctic Sea. As the salts, first gypsum, then common salt, and lastly more soluble potassium and magnesium compounds, crystallized by degrees new water masses entered from the sea. At the same time, the bottom of the gulf slowly receded, giving room for fresh evaporating floods. The salt layers deposited in this manner sometimes reach a depth of more than 1000 m. (about one half a mile). We can thus gain a conception of the immense quantities of water evaporated and the enormous time required therefor. The deposits would long ago have been carried away from their original place were it not for the fact that they finally were covered with a layer of slime nearly impervious to water. The most soluble salts, such as chloride of magnesium, have nevertheless been extracted to a great extent.

The extremes of aridity or humidity have of course not occurred during the brief time known to history. Of a special interest is the question in what direction the climate at present is tending. In this connection, Huntington has aroused great attention by propounding the theory that the earth is now in a period of rapid desiccation.

Judging by the testimony of geology, it seems beyond doubt that an age of humidity prevailed simultaneously with the ice-period in northern Europe, over several parts of the globe, in fact everywhere as far as we know except in Australia. This is clearly borne out by the higher levels of the lakes and their consequent greater extension in earlier days. So far as Tibet and Central Asia are concerned we have already mentioned this fact. But in America and Africa the humid period was even more patent. The Great Salt Lake has covered an area many times greater than at present as testify the picturesque terraces in its surroundings (compare Figure 10). According to the researches by Passage, this period was also strongly pronounced in Africa. A large fresh-water body occupied the Congo basin, the Tsad Lake had a far greater expansion than now, and mighty rivers traversed Sahara.

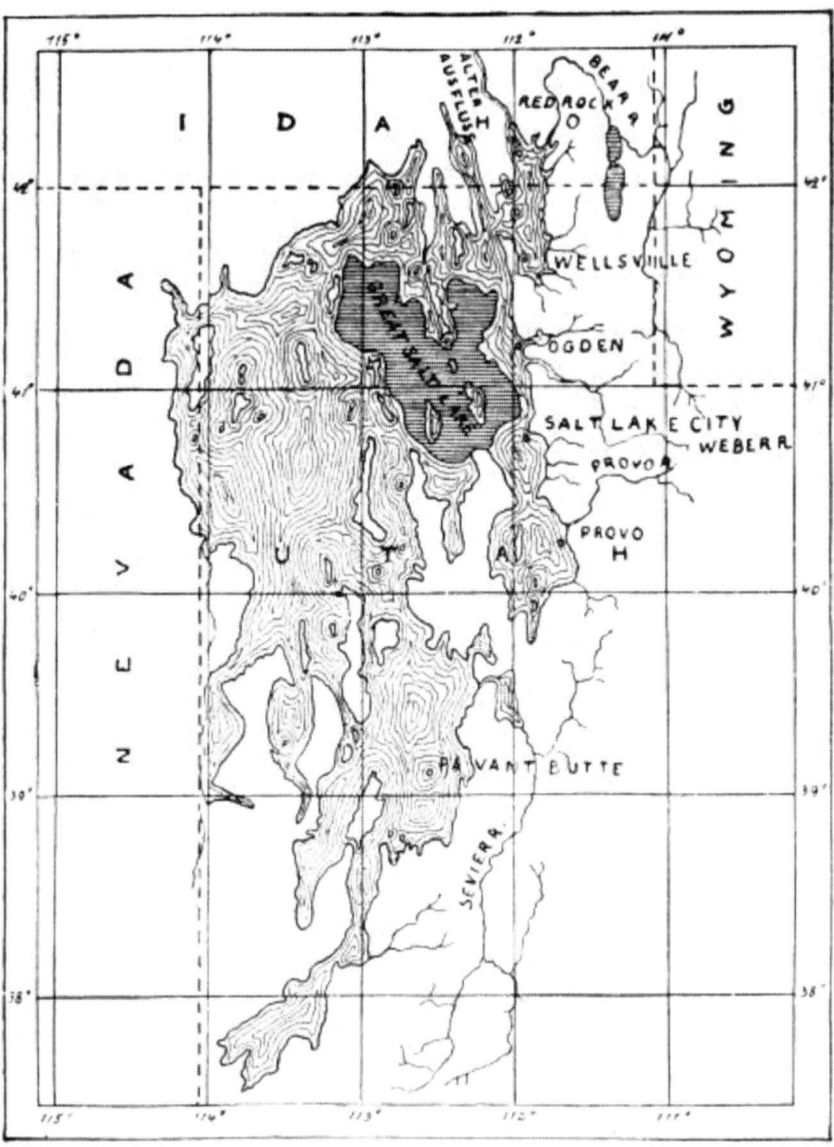

Figure 10. Extension of the great Lake Bonneville in Utah, of which the Great Salt Lake is a remnant.

It is often assumed that the climate of Africa has been more humid even in historic times. The geographer Leo Berg in Petrograd, however, is emphatically opposed to this theory. He points out that the writers of antiquity, Diodoros, Polybios, and Pausanias, have given descriptions of the rivers on the North African coast which nearly agree with conditions today. The location of two ancient cities on the shores of Lake Chott-el-Djerid in Tunis (Lacus Tritonis of old), which lake, it is claimed, reached a level very much higher than now 500 years B.C., plainly demonstrates that the shore-line then ran very close to its present position. Students of ancient Egypt are unable to find evidence of any distinct difference in the climate of that country from the earliest times to our days. It is true that marshes in the Nile delta have changed into splendid meadows—but this is the work of man. The humid period must have ended long before history commenced. A few of the old writers, such as Herodotos, Aristophanes, and Philo, assert that rain never falls in Egypt, but this must be classed as an exaggeration when contrasted with references to rain, snow, and hail in this section made by other ancient authors, for instance Plutarch, Pliny, and Ælianus. At any rate it seems as if precipitation was as rare an occurrence in the land of the Pharaohs as in the land of the Nile of today.

Against the statement by Huntington that the climate of Palestine has become very much more torrid during historic times, stands the assertion by Hilderscheid, who has made a thorough study of these questions, that no reason whatever exists for such a conclusion.

Of greatest interest to us in this connection are perhaps Italy and Greece. Huntington believes that the river Alpheios, which flooded Olympia and covered it with a 4 m. to 5 m. (4 or 5 yards) thick sediment, carried a far greater volume of water formerly than today. This flood, however, was caused by an earthquake accompanied by a fall of rocks whereby the river was dammed. There is no necessity for assuming a greater abundance of the waterflow. According to Strabo, the streams Kefissos and Ilissos, between which Athens is situated, dried out during the summer then as they do now. If we are to believe Pausanias, the brooks which traversed the Argive plain behaved similarly, and so they behave today. From all we can judge, the climate of Greece has not changed perceptibly since the days of Homer.

In regard to Sicily, it is stated that several of its rivers were navigable in the Middle Age, while such is not the case now. But this is explained by the devastation of the forests which formerly equalized the flow of these rivers, and perhaps also by the size of the vessels of that time. Cultivation in these regions has sharply declined since antiquity. As a consequence, the loose soil, which formerly was planted, has been washed away and the dams and retaining walls, built to prevent a too rapid drain of the water, have disappeared. Thus the country has grown increasingly arid. Large cities, as Palmyra, existed in desert regions where lack of water now prevents habitation. But water was brought to the metropolitan cities of old through long magnificent aqueducts, the ruins of which partly remain today. We have all reasons to believe that the marked decrease in cultivation and population, laid to changes in conditions of humidity, depended altogether on man's interference with nature. On certain rocks in Morocco, carvings have been found portraying in a simple way large mammals such as elephants, rhinoceroses, and giraffes, which do not exist in these regions now because of lack of nourishment. But these rough works of art, resembling those of the Bushmen of today, date from prehistoric time, the so-called Paleolithic era, when the climate admittedly was more humid in these regions than it is now.

Similar conditions obtain, according to Hedin, in Central Asia and in Persia. The climate there has without doubt been more humid but not in historic times. The march of Alexander toward India took place under as adverse conditions as now are found in these regions (Baluchistan). Their cities, now in ruins, received their water supply through conduits from rivers some of which were then adjacent to the cities, although they later have changed their course as pointed out by Leo Berg.

In Western and Central Europe numerous marshes and morasses have indeed been drained and rendered available for cultivation, but this does not prove that the climate is become more dry. On the contrary all observations, for instance those made by Tycho Brahe on the island Hven, indicate that the difference between summer and winter temperature has decreased during historic time; that is the climate is become less continental, or more humid, than formerly. Furthermore, many circumstances, such as the occurrence of hazel and of water chestnut in far more northerly latitudes and the higher altitude of the timber line in earlier time, prove that the summer in prehistoric ages was warmer than now. Simultaneously it was dryer. A study of the lacustrine pile dwellings in Switzerland shows that the levels of the lakes then were not higher than now but very nearly the same, which demonstrates that the precipitation has not changed perceptibly in Switzerland since these buildings were made; the period in question occurred, we believe, about 7000 years ago.

While great climatic changes have taken place since man's first appearance on earth, presumably before the end of the ice period, historic time is too short to record any distinct modifications. Local ones may be in evidence such as West Europe's transition to a less continental climate. A variation of this nature has been found not longer back than since thermometrical observations commenced. Thus the winters in Berlin during the period 1746–1847 were colder and the summers warmer than during 1848–1907. The difference for January amounted to -1.5° C. (-2.7° F.) and for May to +0.6° C. (+1.08° F.). The tabulation below, quoted from Ekholm, shows the mean temperature in Stockholm, Lund, London, and Paris, during winter (December-February), spring (March-May), summer (June-August), and autumn (September-November) and for the following periods:

	Stockholm		*Lund*		*London*		*Paris*	
	1799–1848	*1849–1898*	*1753–1798*	*1799–1898*	*1799–1848*	*1849–1898*	*1806–1848*	*1849-1898*
Winter	+25.5° F. -3.6° C.	+26.8° F. -2.9° C.	+30.2° F. -1.0° C.	30.9° F. -0.6° C.	38.5° F. 3.6° C.	39.2° F. 4.0° C.	37.9° F. 3.3° C.	37.9° F. 3.3° C.
Spring	37.9° F. 3.3° C.	37.9° F. 3.3° C.	41.2° F. 5.1° C.	41.5° F. 5.3° C.	48.2° F. 9.0° C.	48° F. 8.9° C.	50.5° F. 10.3° C.	50.3° F. 10.2° C.
Summer	60° F. 15.6° C.	60° F. 15.6° C.	61° F. 16.1° C.	60.2° F. 15.7° C.	61.9° F. 16.6° C.	62.2° F. 16.8° C.	64.6° F. 18.1° C.	64.8° F. 18.2° C.
Autumn	43.9° F. 6.6° C.	43.5° F. 6.4° C.	45.9° F. 7.7° C.	45.9° F. 7.7° C.	50.7° F. 10.4° C.	50.5° F. 10.3° C.	54.2° F. 11.3° C.	51.8° F. 11.0° C.
Year	41.9° F. 5.5° C.	42.1° F. 5.6° C.	44.6° F. 7.0° C.	44.6° F. 7.0° C.	9.8° F. 9.9° C.	50° F. 10.0° C.	51.3° F. 10.7° C.	51.3° F. 10.7° C.

The difference is not great. For Stockholm the winter has grown warmer, the autumn colder, for London the winter warmer and so slightly also the summer, but spring and autumn a trifle colder, and for Paris the summer a little warmer while the autumn is considerably colder. Lund shows the least variation. The winter has grown 0.4° C. (0.72° F.) warmer and the summer colder by the same amount. The annual mean remains nearly constant, only slightly increased, but the climate is become more marine. (This is hardly apparent from the figures cited as far as Paris is concerned.)

From Tycho Brahe's observations of the number of days when snow or rain fell in the place where his observatory was situated on the island Hven in Öresund not far from Copenhagen, Ekholm has calculated that the temperature there during the time 1582–1597 was 1.4° C. (2.5° F.) lower in February and 1° C. (1.8° F.) lower in March than in later years (1881–1896). On the other hand the first autumn frost occurred at the same time as now and the same was the case with the last frost in spring, so that the temperatures on these dates in autumn and spring were nearly identical at the end of the sixteenth century and now. Ekholm drew the conclusion that the climate is become more marine.

Hildebrandson makes the objection that Tycho Brahe's observations were confined to an abnormally cold period judging by the tables Speerschneider has prepared showing the ice formation in Danish navigable waters. Nine out of the sixteen years in which Tycho Brahe gathered his data were notable for abnormally cold winters, while only nineteen out of the hundred years composing the sixteenth century were characterized by equally severe winters.

Thus, the conclusion is not warranted that the winters of the sixteenth century as a whole averaged colder than those of the nineteenth. Later investigations (in 1917) in regard to the dates when the ice would break up in Lake Mälar at Vesterås, in Neva River at Petrograd, and in Dwina River at Riga, have led Ekholm to believe that he has found a periodicity in winter temperatures of not less than 212 years, a conclusion which would agree with Speerschneider's statistics. If so, we are at present living in a period remarkable for its mild winters while a series of extremely severe winter seasons occurred at the time of Tycho Brahe. This law would also have a bearing on the preceding table of temperatures in Stockholm, Lund, London, and Paris as a succession of severe winters came around in the beginning of the nineteenth century while the reverse is true towards its end. On the whole climatic variations during historic time have been insignificant, if present at all, provided we extend our comparisons over two or more centuries. Such is also the opinion of Hildebrandson.

The idea of a slow deterioration of the climate due to increasing desiccation is of old lineage and is most likely related to the venerable conception of a bygone golden age. Aristotle even at that early date believed that a gradual arefaction of the Earth took place. In recent times this faith has been particularly fostered by Huntington in a number of treatises where he endeavours to prove that Asia, represented for instance by Palestine, Syria, and Persia, and further Africa and North America are subject to a rapid exsiccation clearly traceable through historic time. The contrary, however, is true about Western Europe. It was often said that Southern Russia in recent times suffered a slow arefaction manifesting itself in the formation of steppes. This led to careful investigations showing the assertion to be erroneous and culminating in the work of Leo Berg. Rather, a slight shifting in the opposite direction is detectable as the forest region has expanded at the expense of the steppes in conformity with the development toward the end of prehistoric time. The renowned American astronomer, the late Dr. Lowell, embraced the idea of increasing aridity which he observed himself in Arizona where his observatory is located. The drying out of Arizona undoubtedly took place in long bygone prehistoric time. The disappearance of a high culture in Syria and Mesopotamia was the result of hostile devastation of their waterworks; a compensation is now offered in the reclamation of the deserts along the Nile, in California and Arizona, and in numerous other places.

Chapter 4

ATMOSPHERE AND PHYSICS OF THE STELLAR BODIES

In a certain sense, we are justified in speaking of an atmosphere of suns and stars. These bodies consist mainly of a comparatively dense mass surrounded by a layer of very attenuated gas. The density of our Sun is about 1.4 times that of water. In other stars it is considerably lower, in some cases only a few hundredths of the density of water. This applies particularly to those stars of variable magnitude, the Cepheid type, named after their longest and best known representative, the mysterious star Delta in the constellation Cepheus, and in general to young stars. In any case the stars are gaseous throughout on account of their high temperature. An exception must be made for the clouds of matter precipitated by easily condensed vapours, such as gaseous carbon, which clouds float in the outer strata, and are responsible for the bright astral light.

The stars just mentioned belong to the comparatively young stellar bodies, while the Sun in common with other yellow stars is considerably older. Correlated with their age is undoubtedly the greater mean density of the yellow stars. Around many of the young stars, for instance around the brilliant Altair, the principal member of the constellation Aquila, a gas shell of great expansion has been observed, usually consisting of hydrogen but frequently also of helium. These extensive gas appendages may be considered as a kind of atmosphere surrounding the stars in question. Their density is no doubt exceedingly small. Our own central orb, the Sun, is also endowed with rarefied gases outside of the luminous clouds. By their absorption of light they cause the dark lines in the Sun's spectrum, after their discoverer named Fraunhofer lines. The greatest height from the surface of the Sun is reached by hydrogen, mixed with a small quantity of helium and with a gas, unknown on Earth, which we call coronium because it has been observed in the Sun's corona. These gases may be looked upon as the atmosphere of the Sun.

Similar conditions obtain no doubt on the major planets, which possess a density not essentially different from that of the Sun. They have, moreover, practically the same period of revolution around their own axis, Jupiter 9.9 hours, Saturn 10.3, and Uranus (probably) 10.8 hours. Judging by their density, they are, in all probability, like the Sun, gaseous throughout except for the heavy cloud formations which appear to constitute the outer limit of these stellar bodies. Their interior, like that of the Sun, may contain comparatively sluggish gas-masses, as certain peculiar patches appear on their exterior, similar to the sunspots and persisting for long intervals, sometimes over a year. The best known example of the kind is the so-called red spot on Jupiter, which has remained since 1878, although it is not so pronounced now as during its early days (see Figure 11). Characteristic for these planets are certain bands of a marked delineation and running parallel with the equator (see Figures 11 and 12). They are caused by the rapid peripheral motion of these planets, with Jupiter 28 and with Saturn 24 times that of the Earth.

Which gases should we expect to find in the atmosphere of these planets? According to the Kant-Laplace hypothesis, a theory generally credited with a sound kernel, the planets were segregated from the Sun's substance at a time when the latter was expanded so as to include the orbits of these planets and beyond. Naturally, therefore, their atmospheres would originally be composed of the very gases that formed the outmost part of the Sun's atmosphere, notably hydrogen. Slipher, who has photographed the spectra here reproduced of the outer planets, believes that certain strong absorption bands in the spectra of Neptune and Uranus correspond to the distinctive F and C lines of hydrogen, using Fraunhofer's denotation (see Figure 13). But because the bands in question, as shown on the figure, are very broad it is difficult to identify them with certainty. Also other gases of unknown origin enter into the vapour envelopes outside of the clouds and cause, as apparent from their spectra, a strong absorption of the sunlight reflected from the clouds below. The absorption increases with the planet's distance from the Sun; thus, it is most pronounced on Neptune and least on Jupiter.

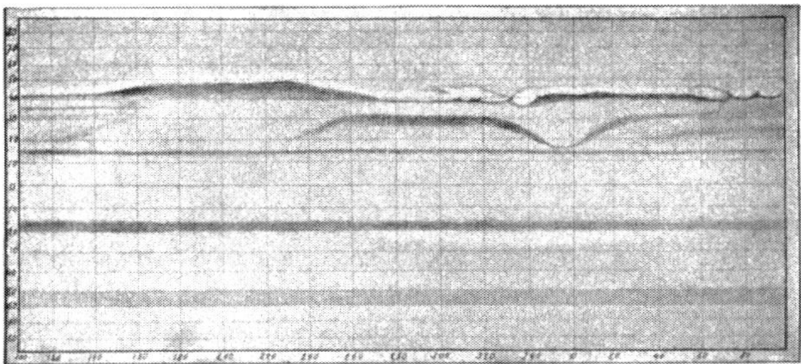

Figure 11. The planet Jupiter in 1909 in Mercator's projection by F. le Coultre of Geneva. The "red spot" over which the clouds are dispersed, is found at 355° Long. and 20° S. Lat. in the outcurving of the dark band. South direction is upward, as on all pictures obtained with astronomical tubes.

Figure 12. The appearance of Saturn September 30, 1909, according to F. le Coultre of Geneva.

At any rate, the gas appendages of the heavenly bodies just considered differ in one essential respect from the atmospheres of the inner planets: Mars, Earth, Venus, and Mercury. On the Sun and on the outer planets the atmosphere gradually merges into the interior gas-masses so that no distinct boundary can be found between the rarer and the denser layers. Widely different conditions obtain on the Earth. Here the range of air is sharply defined below by the Earth's solid crust or by the oceans. In such case alone may we speak of an atmosphere proper, of the kind that enters into our commonplace conceptions. Similar are the conditions on all stellar bodies with a solid or a liquid surface.

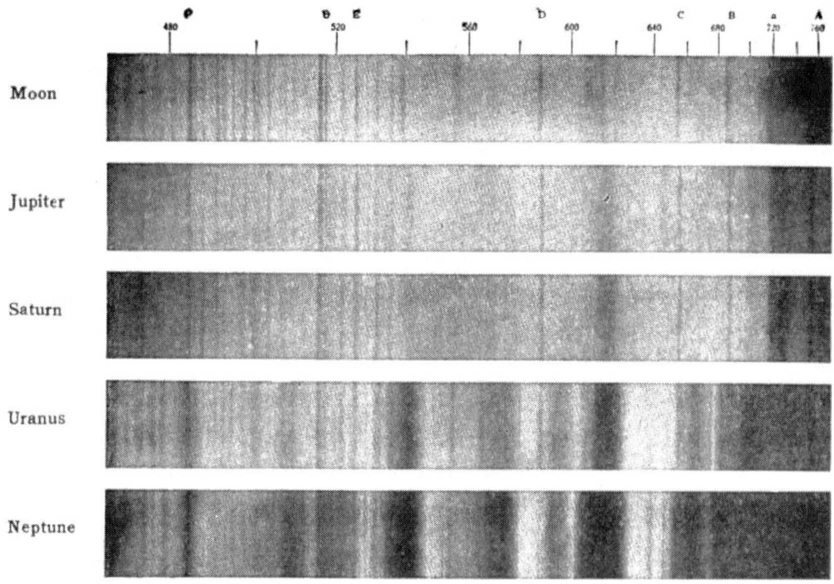

Figure 13. Spectra of the major planets, compared with that of the moon. The latter corresponds to the spectrum of sunlight reflected from a planet which lacks a light-absorbing atmosphere. Photos by V. M. Slipher of the Lowell Observatory.

But it is not certain in every case that all such planets possess an atmosphere. Observations of the Moon when passing some star show that the air envelope, if present, is unable to deflect the light beam from the star, or in other words it has no perceptible power of refraction. From this we also infer that its density is very small, corresponding at most to one or two mm. (.04 to .08 inches) barometric pressure. But we have good reasons to believe that the Moon has been detached from the Earth, carrying away parts of its lightest substance, which theory is supported by the fact that the Moon's mean density (3.3) is only six-tenths of the Earth's (which again is 5.53 times that of water), and we might therefore have expected that the Moon in parting should have shared in the very lightest constituents of the Earth, namely its air-covering. Such was unquestionably the actual procedure, but in the course of time the Moon has lost its originally no doubt considerable atmosphere. The reason is that the molecules in a gas are in a continuous rapid motion, which is the swifter the lighter the gas and the higher its temperature. In hydrogen, the lightest gas known, the velocity amounts to 1.84 km. (1.15 miles) per second at 0° C. (32° F.). The parts of the Moon exposed to the strongest sunlight are heated to about 150° C. (300° F.). At that temperature, the average velocity of hydrogen molecules is 2.29 km. (1.43 miles) per second. But a body departing from the surface of the Moon at a rate of 2 km. (1.24 miles) per second, or more, cannot be retained by the attraction of that globe and therefore never retraces its path but speeds away forever. In the same manner a bullet ejected from a cannon with an initial velocity of 11.2 km. (7 miles)—a velocity not even approached by present artillery—would fly away from Earth barring the resistance of the air. Thus we see that we as yet are far from the realization of the dreams of Jules Verne in his *A Voyage to the Moon*. At any rate, gravity on the Moon is too weak to retain hydrogen over the hottest point of the surface. This part of the gas flies away, new supplies rush in from the sides, and in a short time all traces of hydrogen have disappeared from the Moon. Probably it was mainly gathered in by the Sun, where a velocity of 613 km. (380 miles) per second is necessary if the molecules are to overcome the Sun's attraction, while their actual velocity there amounts to only about 8 km. (5 miles) per second.

In a similar manner we find that the second lightest gas, helium, at a temperature of 150° C. (300° F.), possesses a molecular velocity of 1.62 km. (1.1 miles) per second. This is less than the 2 km. (1.24 miles) per second necessary to leave the Moon's sphere of attraction. But all helium molecules do not move at the same speed; some are faster and some slower than the average. Those moving at a higher rate than 2 km. (1.24 miles) per second constitute a considerable fraction of the total. This fraction disappears. Equilibrium is soon restored so that in less than a second the same fraction of helium molecules is ready to depart. In this manner the Moon lost its helium atmosphere speedily, although not quite as rapidly as its hydrogen.

More slowly yet vanished the gases which are most abundant in our atmosphere, nitrogen and oxygen, but these too were not fettered forever by the limited gravity on the Moon. The same fate befell aqueous vapour, which is nearly twice as light as oxygen. The loss of water, however, was long delayed, as we later shall learn, because new vapour masses were discharged from the lunar volcanoes. In these considerations, we should also bear in mind that the Moon no doubt was a fluid molten mass when separating from the Earth and its substance resembled the lava from our volcanoes. In this condition it remained until its exterior temperature had fallen to about 1200° C. (2200° F.). At that point, the average velocity of oxygen molecules is about 1 km. (.62 mile) per second, with variations in both directions, so that a few percent of them reach a sufficient velocity to leave the Moon forever. Such gas molecules of medium weight return probably to the Earth which, as experience tells us, is ponderous enough to hold them in bonds.

All gases, that constitute any considerable fraction of the Earth's atmosphere, and which, therefore, most likely were divided with the Moon in its parting from us, have again left that globe. The same unquestionably holds true for other stellar bodies of equal or smaller size, such as all the minor planets and for the great majority of the satellites to the major ones. Only the very largest of Jupiter's moons, and possibly Neptune's lonely companion, whose size is not known with certainty, might possibly surpass our Moon in ability to retain gases. Our reasoning with respect to the Moon applies also to Mercury. It is true that the molecules there must possess a velocity one and a half times as high as on the Moon if they are to leave the planet. But at the same time the temperature on Mercury's hottest point, always turned toward the Sun, is far higher, about 400° C. (750° F.), so that the molecules there move 1.26 times as fast as similar molecules over the Moon's hottest point. Mercury is consequently better able than the Moon to retain gases, but the difference is slight. Direct observations (see below) also lead us to believe that Mercury is very similar to the Moon in these respects. We might possibly imagine that certain gases, which on the Moon would condense into fluids or solids, on Mercury might remain volatilized on account of the high temperature and thus form an atmosphere. Such assumption, however, would be erroneous. The investigations by Schiaparelli and by all his successors show that Mercury in turning around its axis always presents the selfsame side to the Sun. The opposite side, never reached by a ray of sunlight, must assume an extremely low temperature, very close to the absolute zero (-273.7° C. or -460.6° F.) and far below any cold existing on the Moon. To this side, all bodies with an appreciable vapour pressure must distil and freeze to solid lumps or frost-coverings without perceptible vapour pressure. For these reasons, Mercury cannot possess any atmosphere to speak of. There remain in the whole series of planets and satellites in our solar system only two bodies besides the Earth which are endowed with an atmosphere in the original sense of the word—namely, Mars and Venus.

We reach the same conclusion when investigating the ability of the planets to reflect the sunlight falling upon them. The bodies which possess an atmosphere hold also suspended therein clouds of water or ice, and also of dust, whirled up from below. These floating particles reflect light far more efficiently than the solid or fluid surface of a planet. The Moon can now reflect 7.3 percent of the sunlight and Mercury 6.9 percent (H. N. Russell, *Proceedings Nat. Acad.*, 1916). These numbers lie so close that they may be considered practically the same within errors of observation.

It is therefore probable that Mercury is as devoid of an atmosphere as the Moon. The opposite extreme is represented by Venus, which reflects not less than 59 percent of the sunlight received, according to H. N. Russell. Terrestrial clouds were found by Abbot to return 65 percent. We believe from astronomical observations that the entire surface of Venus is hidden behind a thick opaque cloud-covering. The slight difference between 0.65 and 0.59 may be due to errors of observation, but also to a small absorption of light in those parts of Venus's atmosphere which are outside of the clouds. Saturn and Jupiter are very similar to Venus in this regard with 63 and 56 percent, respectively. The gases above the clouds on these planets extinguish to a considerable extent the sunlight reflected from the clouds, as apparent from their spectra. (Compare Figure 13.) Hence the value 0.63 given by Russell for Saturn is probably too high. Concerning Jupiter it has been observed that its red light becomes deeper when the sunspots are few, but whiter when the spots are numerous. The sunspots have been found to favour the formation of high clouds, such as cirrus, and this would seem to apply also to Jupiter; when spots are plenty, the clouds are high, and consequently the absorbing layers above, which cause the red colour, are thinner, so that Jupiter will then shine with a whiter—less red—lustre than when the sunspots are rare.

The two outmost planets, Uranus and Neptune, return, according to Russell, 63 and 73 percent respectively of the sunlight received. These figures are probably too high. They do not agree well with Slipher's records of their spectra (Figure 12).

There now remains Mars. This planet approaches the Moon inasmuch as it reflects only 15.4 percent of the sunlight arriving to the orb. Everything points to the conclusion that the atmosphere of Mars is very thin. Lowell estimates, on somewhat meagre grounds, however, that on each square metre of the planet rests only 22 percent of the mass of air supported by each square metre of the Earth's surface.

It would naturally be very interesting to ascertain the amount of sunlight our Earth throws back into space. This we cannot measure, as we cannot place our instruments outside of the Earth's cloud-mists nor can we read them there. Not less than 52 percent of the Earth is covered with clouds, whose whiteness (Latin: *Albedo*) is 65. Thus the clouds alone return $0.52 \times 0.65 = 0.338$ parts of the sunlight. Of this portion a fraction amounting to about 4 percent is extinguished in the air above. The remainder is 0.325. Atmosphere and suspended dust reduce the sunlight over the cloud-free part, *i.e.* 48 percent of the Earth, by 60 percent, half of which returns to space, while the other half reaches the ground in the form of light from the sky, and of this fraction again about 4 percent is reflected into space; these two items added give 0.15. Finally, the 40 percent sunlight directly received by the Earth's surface is reflected to the extent of 6 percent by the oceans and by the generally moist ground; deserts and bare rocks reflect about twice as much, but their total area is comparatively small; of this 6 percent reflected light, 70 percent reaches outside space; thus we obtain $0.48 \times 0.40 \times 0.06 \times 0.70 = 0.008$. In all, therefore, the amount of reflected sunlight is $0.338 + 0.15 + 0.008 = 49.6$ percent If the air were free from clouds, the reflexion-number or Albedo would be 33 percent, or considerably higher than that of Mars. When now half or a little more (52%) of the Earth's surface is overcast with clouds and this portion therefore has the whiteness of Venus, the figure 49.6 (Russell calculates the figure 45) for the entire Earth naturally falls closer—almost 3.6 times—to 59, the figure of Venus, than to 15.4, the figure for Mars. We may also compare the value 33 percent, which applies to the cloud-free portion of the Earth, with the value 15.4 percent for Mars, which is almost without clouds, and with the value 7.3 percent for the Moon, which has neither clouds nor dust, because it lacks an atmosphere. We can then conclude that the atmosphere of our Earth holds almost three times as much dust suspended over each square metre as does Mars, and this in spite of the smaller gravitational force on Mars, which is about 37.5 percent of that on Earth. Taking proper account of the low temperature on Mars we may easily compute, by means of a formula given by Stokes, that a particle of dust should sink 2.3 times slower on Mars than on Earth. When, nevertheless and in spite of frequent but thin mists, so few particles of dust float in the atmosphere of Mars, the conception inevitably comes to our mind that the air on that planet must be extremely rarefied so that the wind-puffs have little power to raise the dust from the ground. Lowell estimated the barometric pressure at the surface of Mars to be about 64 mm. (2.52 inches), and Proctor gives about twice this figure. There appears to be ground for considering already the former value too high; both are very uncertain. If we accept that of Lowell, we find that each square metre of the surface of Mars supports a column of air, whose mass is only about one-fifth of the mass resting on each square metre of ocean surface on Earth.

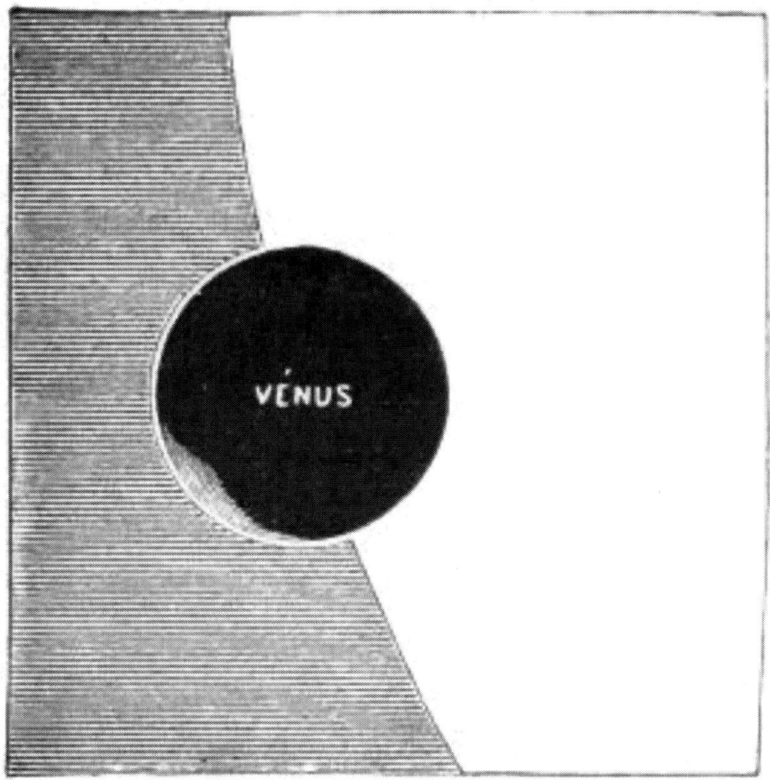

Figure 14. The planet Venus, with sunlit atmosphere (to the left), as observed by Langley at the Venus passage December 6, 1882.

The dense clouds which float above Venus have long ago led to the assumption that the atmosphere of that planet must be far deeper than that of the Earth. Its strong refractory power has also contributed to this belief. When Venus is close to the sun-disc the dark body stands forth surrounded by a ring of light (see Figure 14). It is, however, recognized that this phenomenon requires no greater air density than that on Earth for its appearance. In this connection we should remember that the inside limit of the vapour shell which we in this manner observe, is the cloud-wall, not the ground. And these clouds, we have every reason to believe, float on account of the heat prevailing at a great height in the atmosphere, so high in fact that they form an impenetrable wall already where the cirrus clouds appear in our sky. If these suppositions are correct, the light-ring mentioned is caused by a quarter only of the air-masses on Venus, and its total air-covering must be far deeper than that of the Earth. The latter occupies probably in this respect as well as in reference to position in space a middle ground between Mars with its extremely thin and Venus with its comparatively dense atmosphere. If so, we might expect the atmosphere on Mercury to be denser yet, while we already have seen that it is almost wholly lacking on that planet. The explanation is that Mercury has lost its spin around its own axis and therefore always presents the same side to the Sun—just as the Moon and probably all other satellites turn one side only toward their respective central bodies—hence the opposite side becomes so cold that all gases are there condensed to fluids or solids except the two most volatile ones, hydrogen and helium, which on the other hand leave the planet on its hot side. If Venus, therefore, as held by several astronomers from Schiaparelli to Lowell, always turned one side only toward the Sun, this planet also would be without any perceptible air-covering. According to investigations by Bjelopolsky of Venus's spectrum, which investigations, however, are in complete disagreement with corresponding measurements by Slipher, that planet has a period of rotation on its own axis of about 29 hours. This figure is very uncertain and a new determination is therefore highly desirable.

In order to understand the atmospheres of the planets, it is of great interest to ascertain the composition of the air that surrounds the Earth. Our knowledge in these matters has grown considerably of late. We shall in the main follow the presentation by Dr. Wegener of Marburg.

We know at present with considerable accuracy which gases enter into the air. Besides the previously well-known nitrogen and oxygen which contribute the bulk, 78.1 and 20.9 percent respectively, of the total volume at the earth's surface, we find water vapour in proportions changing with localities and times, and it is for this reason left out when fixing the various percentages; further, carbon dioxide 0.03 volume percent and the rare gases discovered by Rayleigh and Ramsay, argon, 0.932 percent, neon 0.0012 percent, helium 0.0004 percent, krypton 0.000005 percent, and xenon 0.0000006 percent Each one of these constituents diminishes in quantity with height in accordance with the so-called barometer-formula and the rate is the more rapid the heavier the gas. Krypton and xenon, therefore, which are two and one half and four times heavier than oxygen, occur mainly in the lower strata. The percentage of helium on the other hand, a gas eight times lighter than oxygen, should increase rapidly with height. If the air consisted of a mixture of oxygen and helium at 0° C. (32° F.) the former would decrease to one half at an elevation of 5 km. (3.1 miles) but the latter not before we had ascended 40 km. (25 miles) (eight times higher than for oxygen, as the weights are in the ratio of 1 to 8). At that altitude, oxygen would have decreased in the proportion $1:2^8 = 1:256$. When, as actually is the case, there is 50,000 times more oxygen than helium at the surface this ratio should decrease in the proportion 128:1 at a height of 40 km. (25 miles). Ninety kilometers (56 miles) above the surface helium should overbalance oxygen and thereafter rapidly gain in preponderance. This holds true provided no agitation takes place in the form of vertical currents of air.

Similar laws apply to all light gases which do not turn into fluids or solids at low temperatures. Aqueous vapour, on the other hand, which when cooled condenses to clouds, diminishes much faster than the nearly twice as heavy oxygen, because the temperature rapidly decreases as we move upward or at a rate of about 5° C. per km. (14.5° F. per mile) up to 2.5 km. (1.5 miles) and of 8° C. per km. (23° F. per mile) at a height of 8.5 km. (5.3 miles). The quantity of water vapour shrinks to one half at 1.9 km. (slightly more than a mile) above ground. Carbon dioxide again follows the barometer-formula applicable to other gases because it occurs in such minute quantity that it never condenses to clouds. In fact, it is water vapour alone which must be treated as an exception. Carbon dioxide is nearly one and one half times heavier than the other gases of the atmosphere on an average. It should therefore diminish in the proportion $1:2^{1.5} = 1:2.8$ in a vertical distance of 5 km. (3.1 miles) while the density of the air decreases only in the ratio 1:2. Several determinations of the presence of carbon dioxide in the atmosphere as high up as 3.8 km. (2.33 miles) have been made, by S. A. Andree among others, but the percentage of this gas remains constant within the errors of observation. The same holds true to a height of 7 km. (4.35 miles) for the proportion between oxygen and nitrogen, although we might have expected a perceptible change as oxygen is 14 percent heavier than nitrogen. How shall we explain this fact which seemingly contradicts the theory just advanced?

The explanation is quite simple. The preceding statements hold true for a mass of air at perfect rest. But, if the air is violently agitated, the composition becomes homogeneous all through. We know that in the barometric cyclones and anticyclones strong rising and descending air currents flow. The composition of the atmosphere, therefore, becomes the same as far up as this mixing action prevails. These currents produce another effect, namely, a fall of temperature with rising height. Because when a gas is transported upward the surrounding pressure decreases, resulting in expansion and consequent cooling. It is well known that a gas is heated when (rapidly) compressed, a quality formerly made use of in the pneumatic fire-tool to ignite tinder. It is evident that conversely a gas must cool off when expanding. If now the mixing of the air were extremely rapid the thermometer would fall very close to 10° C. (18° F.) with each km. (.62 miles) rise in elevation. If, on the other hand, the air stood perfectly still in a vertical direction, the temperature would remain constant at all heights over the same point. Between these two extremes, we find the actual condition, inasmuch as the temperature of the atmosphere decreases upward 5° to 8° C. per km. (14.5° to 23° F. per mile) as observed during balloon ascensions.

This applies to the so-called "troposphere"—mixing-zone. One of the most remarkable discoveries in recent times, made by Teisserenc de Bort and Assman, is the fact that the decrease of temperature with height does not continue indefinitely but only up to a certain elevation,—in middle Europe about 11 km. (7 miles), in Lapland about 7 km. (4.5 miles), and at the equator about 15 km. (10.5 miles)—and above this point the temperature remains constant. We now meet the peculiar condition that the temperature of this upper layer, which is called stratosphere—"film-zone"—is lowest over the equator, because it commences at a great height there, and lowest over the polar regions, where it extends farther down. The stratosphere has received its name from the fact that it consists so to speak of lamellæ almost parallel to the earth's surface and moving in a horizontal direction while vertical motions are absent. The winds in these strata have a marked westerly direction (*i. e.* they are east winds) and they become stronger the higher the stratum—at an altitude of 83 km. (52.5 miles) their velocity is about 100 m. (330 ft.) per second. In the troposphere on the other hand west winds predominate. The wind direction in the stratosphere was observed on the so-called luminous night-clouds which were found as high as 80 km. (50 miles) above the Earth. These strata consequently revolve slower around the earth's axis than the solid body of the planet itself. At an elevation of 80 km. (50 miles) the rotational speed has decreased to 65 percent of the angular velocity of the earth's surface. We have reason to believe that the very highest strata stand still, that is do not take part in the earth's rotation on its axis. This would follow if outside space were not entirely devoid of vapour so that our atmosphere would merge imperceptibly into the exceedingly attenuated gas masses of interplanetary space.

As high as the mixing-zone extends, so high also is the composition of the air constant and like that at the surface of the earth. But above this limit—in Scandinavia, we might say above an elevation of 10 km. (6.2 miles)—commences a rapid tapering of the heavy gases, while the percentage of the light ones correspondingly rises. Foremost among the latter is hydrogen, with only half the weight of helium. The presence of hydrogen in the atmosphere has been shown by Boussingault, and the proportion in which it occurs has later been measured by Armand Gautier. It is about one three hundredth part of one percent. It increases extremely rapidly with height in the stratosphere so that 80 km. (50 miles) above the earth and upward hydrogen is more abundant than all other known gases of the atmosphere at the same altitudes.

We reproduce below a somewhat revised table by Dr. Wegener of Marburg, who has made the most recent computation of the percentages of the various constituents of air at different heights. Consideration has been given to the fact that the composition of the air does not change except as regards the percentage of moisture within the troposphere, which is assumed to reach a height of 10 km. (6.2 miles). As usual in similar cases the percentages refer to volume.

Height in km.	Height in miles	Pressure in mm.	Pressure in inches	Hydrogen 2	Helium 4	Nitrogen 28	Oxygen 32	Argon 39.9	Carbon Dioxide 44	Water 18
0	0	760	29.9	0.0033	0.0005	78.1	20.9	0.937	0.03	1.41
10	6.2	197	7.75	0.0033	0.0005	78.1	20.9	0.937	0.03	0.14
30	18.6	8.95	.352	—	—	85	15	0.29	0.0064	0.5
50	31.0	0.45	.0177	1	—	88	10	0.10	0.0014	1.7
70	43.5	0.045	.00177	13	1	80	6	0.05	0.0005	—
90	55.8	0.0157	.00062	68	5	26	1	—	—	—
110	68.2	0.0116	.00046	94	5	1	0	—	—	—
130	80.6	0.0097	.00038	96	4	0	—	—	—	—
210	130.2	0.0055	.00022	99	1	—	—	—	—	—
310	192.6	0.0032	.00013	100	—	—	—	—	—	—
410	254.2	0.0021	.00008	100	—	—	—	—	—	—
510	316.2	0.0016	.00006	100	—	—	—	—	—	—

Under the name of each gas its molecular weight is given as a measure of corresponding specific weight. The quantity of water vapour was not included when the percentages of the other gases were calculated, because it changes considerably with locality and time. The number given in the table for water is the mean for the entire globe—it corresponds to 11.4 grammes per cubic metre (.31 oz. per cu. yd.)—or the amount present in air saturated with moisture at 16.5° C. (61.7° F.). The bulk of the water vapour forms a layer strongly concentrated toward the surface of the earth. Carbon dioxide also tapers rapidly with increasing height because its density is 1.5 times greater than that of air. This is apparent from the molecular weight 44 stated under carbon dioxide, while the average molecular weight of air is 29. Faster yet do krypton, with a molecular weight of 83, and xenon, with a molecular weight of 131, decrease as we ascend in the atmosphere. These gases, like neon, whose percentage first increases slightly with height, and argon, which decreases upward as shown in the table, do not perceptibly influence the processes of nature. The reverse is true about water vapour and carbon dioxide, which nourish the plants and also protect the Earth against a too rapid heat radiation into space. We well remember how abruptly the temperature changes in the course of the day in the dry desert climate, while corresponding variation is comparatively slight in humid climates. This is the result of the ability of water vapour to arrest the radiation from the Earth. Carbon dioxide is about evenly distributed over the globe—although somewhat sparser over highlands—and its heat-conserving and equalizing influence is, therefore, not so manifest as that of moisture. Only by the most accurate investigations has this influence been demonstrated.

In Wegener's table a gas is included, called Geocoronium, whose existence in the air has not been directly proved. Conspicuous is, however, the green light displayed at great altitude by the Northern Light arches, a green color which does not, as far as we are aware, belong to any known constituent of the air. It is true that the corresponding spectral line (557 $\mu\mu$) lies very close to a line belonging to krypton, but the latter is a heavy gas which cannot occur to any traceable extent in the high strata, more than 300 km. (186 miles) above the earth, where occasionally the Northern Light arches appear—their favoured height according to measurements by Störmer is about 120 km. (75 miles). Wegener assumes, therefore, that this green line belongs to a hitherto unknown substance, Geocoronium, which should be five times lighter than hydrogen. Recent researches present great difficulties to the acceptance of this assumption, and for this reason further discussion of the problem will be omitted. Above a height of 210 km. (130 miles), this gas, according to Wegener, would preponderate. If such postulated gas does not exist, hydrogen completely dominates in these regions and down to 85 km. (53 miles) above the Earth. Because hydrogen is so light, the density of the air in the range of a barometric pressure below 0.02 mm. (.0008 inches) increases but slowly as we descend toward the Earth. This uppermost part of the atmosphere may appropriately be designated as the hydrogen-zone. Even within this range, the "shooting stars" meet sufficient resistance to flame into light at a height of about 120 km. (75 miles) and dissolve into dust which turns dark about 85 km. (53 miles) above the Earth. E. C. Pickering recognized the spectrum of hydrogen in the light of meteors passing at great height, but decomposed water vapour might possibly be its source. Meteors crossing lower strata show the spectrum of nitrogen. Nitrogen becomes important from a height of about 85 km. (53 miles) downward and from 75 km. (46.5 miles) to the surface of the Earth it predominates. As a consequence the pressure increases rapidly as we approach the ground. In these regions or up to 80 km. (50 miles) floated the highest luminous night clouds, observed by Jesse, indicating that here commenced a new range, the nitrogen-zone. Only the heavy meteorites are able to penetrate into the nitrogen sphere, which checks their speed and causes them to explode, and thereafter the remnants fall with a velocity compatible with the air resistance they meet. To these parts descend also the lowest rays of the Northern Lights, the so-called draperies—Störmer observed them once at a height of 37 km. (23 miles). Finally, water vapour presents itself in appreciable quantities at an elevation of about 10 km. (6.2 miles), where the troposphere commences. We now meet the highest clouds, cirrus (with the exception of the "luminous night clouds" observed only in the years 1883–1892 after the eruption on Krakatoa). To these heights, reach the vertical air currents which are essential to cloud formations. Only light clouds, however, float at these altitudes; the heavy clouds (alto-cumulus) do not rise above 4 to 5 km. (2.5 to 3.1 miles) and the rain clouds proper (cumuli) occur only below a height of 2 km. (1.25 miles). This is the result of the downward concentration of water vapour within the troposphere.

If gravity decreased in intensity, the effect would be the same as if the gases were lighter. On Venus, the intensity of gravity is eight tenths of that on Earth. The difference is slight. If everything else were similar the various air-zones would reach 25 percent higher on Venus than they do on our globe. But one essential condition is varied by the far higher temperature on our neighbour. The proportion of moisture in the air is thereby vastly increased. The dense clouds rise to much greater heights than on Earth. If there be ten times as much water in the air on Venus as there is in the air on Earth—which might fairly represent the actual condition—the heavy rainclouds would there rise to a height of more than 10 km. (6.2 miles), and their smaller weight on Venus would also contribute to their buoyancy. The light cirrus clouds should appear as high as about 30 km. (18.5 miles) above the ground. Under such circumstances, we cannot expect but that the planet must be entirely hidden from our sight as well as from the rays of the Sun.

On Mars, the intensity of gravity is 2.68 times smaller than on Earth. In consequence barometric pressure falls 2.68 times slower with increasing height there than here. The same ratio holds for decrease in temperature and for shrinkage of proportion of moisture when comparing conditions on the two planets. The strong cold precludes anything but insignificant quantities of water vapour. The air on Mars is similar to the atmosphere on Earth in and above the cirrus-region. The clouds existing there are not only extremely thin and transparent—it is well known that cirrus clouds throw no shadows—but they are confined to small fractions of the planet's sky. They are replaced by light mists.

We shall later return to the consequences of these peculiarities.

With help of the spectroscope we have ascertained that the gases on the Sun are, in the main, also arranged according to specific weights, so that the lightest reach the greatest heights. Somewhat similar conditions obtain in the gas-appendages of the stars.

Chapter 5

THE CHEMISTRY OF THE ATMOSPHERE

Of a very particular interest is the question of the atmosphere of the planets. The great problem of habitability of the latter is most intimately connected therewith. Primitive fancy, very early, populated the stellar bodies, especially the stars and the Sun, with beings similar to those on Earth. Gradually, however, the insight awoke that these bodies are incandescent and therefore unfit to shelter life of the kind with which we are familiar. Attention then turned to the planets, as they and the Earth belonged to the same order of heavenly bodies. Perhaps they furnished abodes for our kin. And the stars, suns like our Sun, should not they be surrounded each with its throng of planets, gravitating around their central source of light and heat? This beautiful thought vied with the conception of Earth as the centre of the universe and as wholly set apart from the other stellar bodies, whose prime object it were merely to furnish light and to indicate time for the inhabitants of the Earth. Much to be regretted, it was the latter far less attractive theory which gained firm hold on the Church, although a few of its unbiased men, like the renowned Cardinal Nicolaus Cusanus (1401–1464), declared in favour of the contrary opinion, and did so unmolested. But times grew harder, ironclad orthodoxy triumphed, and Giordano Bruno, whose defence was that he simply accepted the theory of the great Cusanus, was burned at the stake to expiate his fearless assertion that other worlds, no less than ours, might be blessed with the presence of living beings.

Undoubtedly, other planets are upbuilt of the same material which enters into the composition of the Earth—as held already by Leonardo da Vinci. Spectral analysis teaches us that the same constituents form all the suns, including our own. If we agree that the suns have furnished the original substance of the planets revolving around them it is a natural conclusion that this matter should combine into similar chemical compounds on planets equally advanced in their evolutionary, that is cooling, process. And we know, indeed, that the same elements compose the Sun and the Earth, and that the samples brought to us from other worlds, *i. e.* the meteors, are of a composition which strongly reminds us of the rocks in the interior of the Earth. We seek in vain only for indications of the action of water, which substance has left such obvious traces on the surface of our globe and in the immediate strata below. But it should be remembered that the water, in the form of vapour, as previously set forth, has left all the minor stellar bodies, and the meteorites manifestly belong to the small or smallest among the wanderers of the heavens.

There is then no reason to doubt that the material of which the planets are built essentially is the same throughout the universe. Their interior should, like that of our Earth, consist of the heavy metals, principally iron,—strongly prevalent also in the Sun and in the meteors,—and this metallic nucleus should be clothed with the silicates, oxides, carbonates, sulphides, and hydrates of all metals, particularly aluminum, and among the metals we may also count hydrogen. The melting points of these exterior and lighter substances lies above 1000° C (1800° F.). No life could exist in such a molten mass, so that not until a solid crust had been formed through cooling was the possibility of life at hand.

Life is, at least on Earth, tied to certain so-called compounds, in which carbon is the essential common element, while hydrogen, nitrogen, and oxygen together with sulphur, phosphorus, iron, magnesium, and a few other less important elements also enter therein. No substance but carbon possesses this quality of being a prerequisite of life. Silica is a close kin to carbon and a substitute in certain organic compounds, but protoplasm, the main constituent of the living cell, cannot be built without carbon. In the inorganic world, however, silica by virtue of its affinity, which is kindred to that of carbon, plays a rôle somewhat similar to the latter in the almost infinitely variably silicates. Protoplasm cannot endure in a temperature above 60° C. (140° F.) or thereabout—certain algæ, it is sometimes stated, thrive in hot springs up to 80° to 90° C. (176° to 194° F.) but certainly not over 100° C. (212° F.). At these temperatures—strictly speaking at all temperatures between O° C. (32° F.) and 365° C. (689° F.)—water can exist in fluid state and this too is a prerequisite of life. We may say therefore that life is confined to a small temperature range between the freezing and boiling points of water. But wherever water occurs, except in a vessel which it completely fills, there exists also in the adjacent space, if unoccupied by fluids or solids, water vapour of at least 4.6 mm. (.18 inches) pressure. There is, therefore, always an atmosphere of aqueous vapour on any planet whose surface is partly covered by water.

The palæontologists have agreed that all life commenced in the water. The manifold living beings which now inhabit the solid crust of the Earth all descend from ancestors which floated in the waves of the ocean, the cradle of all organisms. It is not absolutely certain that oxygen is necessary to all living beings but many biologists hold that opinion. Certain bacteria are able to draw the oxygen they require for their development from compounds in which oxygen is bound sometimes in a very intimate manner, as in sulphates. But these bacteria are considered degenerate plants, and free oxygen is certainly indispensable to the existence of the animals and probably also to the plants with the exception just mentioned. As we shall see later, free oxygen cannot be present on the planets until a solid crust has been formed. We may therefore state that the conditions for the existence of life on a planet are fulfilled when a true atmosphere containing oxygen and water surround its body.

If we are to understand these conditions, we must study the processes whereby oxygen is supplied to the atmosphere. As the planets are segregations from the Sun, they should originally have the composition of the Sun, particularly that of its outer layers. Here metals occur in greatest abundance, but there are also a few oxides, especially those of titanium and magnesium (according to Fowler), hydrogen in great quantities, oxygen, carbon, cyanogen, and carbon monoxide. It may seem strange that free oxygen exists side by side with a surplus of hydrogen and sodium, strong so-called reducing substances which bind oxygen. But at the high temperatures prevailing on the Sun compounds of oxygen and the reducing substances, for instance hydrogen, *i.e.* water, are largely dissociated into their constituents. But, if the temperature should drop to about 1200° C. (2200° F.), at which point crust building does not yet take place, the oxygen would be entirely absorbed in the formation of the compounds mentioned. The compound substances of the Earth, like those of the Sun, are also strongly reducing, so that we must infer that free oxygen did not enter into the gas covering of the Earth at the time when its solid crust was formed. We may gain a conception of the gases which then existed in the outmost layers of the Earth by studying the gases on the Sun and on other stellar bodies, particularly the comets, and also by investigating the gases absorbed by the molten interior of the Earth. Previous to the crust formation, the entire mass of the Earth except the gases in its outmost layers were of the same character as possessed by its molten interior now. This molten mass, or magma, when in contact with the surrounding gases partly imbibed them through the process called absorption. An investigation of the gases present in the magma will therefore give us an idea of those existing in the original vapours surrounding the Earth. The magma occasionally comes into view in volcanic eruptions, and the confined gases are then partly given up into the air, but they are also partly retained in the solidifying lava, or volcanic rocks, whence they may be driven off by high temperature and subsequently analyzed. The direct gas emanations from the craters may also be gathered and analyzed. Such investigations have been carried out on a large scale by Albert Brun, Frenchman, Arthur Day, American, and his co-labourers Shepherd and Perret. Brun reached the surprising conclusion that water vapour, hitherto considered the most important of the volcanic gases, in reality was not one of them but originated in the crust of the earth. This theory, however, was completely refuted through the investigations by Day and his associates.

As an example, we give the analysis (the mean of several determinations) of volcanic gases from the crater Halemaumau on the volcano Kilauea in Hawaii:

May, 1912 percent of Volume	
Carbon dioxide	55.4
Carbon monoxide	4.3
Hydrogen	7.7
Nitrogen	29.6
Sulphurous acid	2.9
December, 1912	
percent of Weight	
Carbon dioxide	42.9
Nitrogen	25.8
Water	27.5
Sulphurous acid	3.7

In the latter case, it has been shown that air had gained entrance to the volcanic gases, which air might have carried a quantity of water. But this quantity could not have been large judging by the amount of nitrogen present which corresponds to not quite 3 percent water by weight. In the former case water was not included in the analysis. At any event, a high percentage of water has frequently been observed in volcanic gases.

When the gases were left in contact with water a considerable part was absorbed thereby, particularly compounds of chlorine and fluorine and also ammonia and sulphurous acid. An analysis of such water showed 10 percent more fluorine than sulphurous acid and two fifths as much chlorine as fluorine. The ammonia amounted only to half of one percent of the chlorine. None of the rare air gases were present, which indicates that the nitrogen originated entirely from the magma and not from the air.

Brun has analyzed lavas from different volcanoes. The gases extracted therefrom naturally do not give as reliable information about the original atmosphere of the Earth as do the gases directly emanating from the volcanoes. As examples we quote the composition of the vapours in lavas ejected March 4, 1901, from Stromboli, and from Vesuvius in the well-known eruption of 1906. They show in percentages of volume:

	Stromboli	*Vesuvius*
Free chlorine	12.8	0
Hydrate of chlorine	2.0	6.6
Sulphurous acid	4.5	12.0
Carbon dioxide	60.2	73.8
Carbon monoxide	11.5	traces
Hydrogen	0.5	7.6
Nitrogen	6.9	traces
Marsh-gas	1.6	0
	100.0	100.0

It will be noticed that the composition varies considerably. Free chlorine cannot very well be primigenial as it, like oxygen, combines with reducing substances. Chlorine may be produced by heating chloride of calcium with silica and ferro-silicates which are present in the magma. At all events, carbon dioxide constitutes the bulk. Next in importance are sulphurous acid and hydrate of chlorine. Carbon monoxide, hydrogen, and nitrogen may occur in quite considerable quantities but are sometimes almost wholly absent.

Day and Shepherd reached the conclusion that the gases emitted by the hot lava in the Halemaumau crater are nitrogen, water, carbon dioxide, carbon monoxide, sulphurous acid, hydrogen, sulphur vapour, also small quantities of chlorides, fluorides, and possibly ammonia. Such, at least approximately, should also the original composition of the Earth's atmosphere have been when the crust was newly formed. Nitrogen, water, and carbon dioxide were the most important ingredients; in the high strata hydrogen was present. Oxygen was totally lacking and reducing gases such as hydrogen, sulphurous acid, and carbon monoxide abounded instead. If we further note the composition of comets and meteorites we find that cyanogen, carbohydrates, and carbon monoxide also are present in the former, argon and helium in the latter. It is therefore probable that these substances, although absent in the emanations from Kilauea, yet belonged to the primary atmospheres of the planets. The rare gases of the air especially should originally have come from the exterior parts of the sun, as did the nitrogen.

An atmosphere of such a composition would be utterly unendurable to living beings. It must, if organisms are going to thrive therein, be purified from such poisons as carbon dioxide, gaseous sulphur, cyanogen, and sulphurous acid. We know that such a process has taken place and that the sunlight has been the great chemist who produced oxygen and carbon from the carbon dioxide. The just mentioned poisonous gases were subsequently oxidized through electrical discharges. We all know that the plants up-build their framework under the influence of sunlight, consuming in the meantime carbon dioxide, water, and a little ammonia. In the process, oxygen is formed, as are also starch, cellulose, sugars, and albuminous substances with the aid of the green colouring matter in plants, chlorophyll, which considerably accelerates the action. Subsequently these new substances, which all (except the albuminous ones) belong to the carbohydrate group, are converted into principally carbon and water. The final result is that carbon dioxide through the agency of sunlight is split into its two constituents, carbon and oxygen. This process, which is comparatively rapid in the presence of chlorophyll, should also, although more slowly, take place without that medium; and in recent trials the chemists—particularly Daniel Berthelot—have, indeed succeeded in imitating without chlorophyll this important function of the plants through the application of light of a short wave length. In the course of the many millions of years which geology has proved necessary for the evolution of our planet, the carbon dioxide in the air was gradually converted into oxygen and carbon. As long as reducing gases, such as the poisonous ones mentioned, or any considerable quantities of carbohydrates and hydrogen yet remained in the atmosphere, the oxygen was consumed in their combustion. If a solid crust had not existed to prevent the oxygen from entering into the interior molten mass, it would also have found its way there and would have oxidized the reducing substances in the magma. The separation of the interior from the surrounding gas shell is therefore a necessary requisite for the existence of free oxygen in the air. Another condition is that the combustible gases escaping from the volcanoes must be added to the air at a sufficiently slow rate not to consume all of the simultaneously formed oxygen. A third requirement is that the liberated carbon should not in a renewed process of oxidation bind the oxygen just recovered. As long as the air still was reducing, this last condition was no doubt fulfilled for that very reason. At all events, the crust once formed and the original violent volcanic activity somewhat abated, the time finally arrived when free oxygen existed in the air. The previously present reducing gases were, except for small fractions, burned into water, carbon dioxide, and sulphuric acid and the nitrogen compounds had no doubt yielded free nitrogen to the stores of that gas already a part of the atmosphere. The time was now ripe for the first plants, probably low forms of algæ, which, in the oceans, commenced life on our planet. The carbon dioxide and hydrochloric acid of the air as well as the newly formed sulphuric acid were absorbed by the running water and caused a rapid disintegration producing silica and acid silicates. As plant life developed and extended the formation of oxygen increased. Falling vegetable matter was imbedded in slime which prevented the access of oxygen during decay and in this manner the fossil fuels were deposited. Koene of Brussels first pointed out that the carbon and the sulphuric compounds accumulated in the Earth would suffice to bind the oxygen of the air. Later investigations lead to the conclusion that the carbon alone is sufficient for the purpose. It would therefore appear that all the oxygen of the air is derived from carbon dioxide belonging to the original atmosphere or contributed thereto by the volcanoes.

The reason why carbon dioxide and water continuously are liberated from the magma is undoubtedly that the acid silicates are lighter than the basic and therefore accumulate in the exterior parts of the magma. A great surplus of silica exists there. Compounds containing water and carbon dioxide, *i.e.* hydrates and carbonates, are also light and should therefore congregate to the same strata where silica abounds, there partly to be dissolved by the free silica and thus setting water and carbonic acid free. The latter are, in contrast to silica, volatile and evaporate therefore into the air leaving the silica behind. This process is yet in evidence wherever the magma emerges as through the volcanoes. But also a few other acids in the magma are highly volatile, as sulphurous, thiosulphuric, and hydrochloric acids. These also belong to the volcanic gases, are dissolved by the water, and partake in the processes of disintegration. The carbon dioxide and hydrochloric acid form carbonates and chlorides. The former are extracted from the sea water by crustacea, sometimes also by plants, and form part of our sedimentary strata; the latter are soluble and remain in the water, chiefly as sodium chloride or common salt. The thiosulphuric acid, probably a product of ferrous sulphide and acids in the magma, has entered into numerous insoluble metallic sulphides found in the Earth. Partly, it has also been oxidized, like the sulphurous acid, into sulphuric acid and has then assisted in the processes of disintegration, forming gypsum which has been deposited in the sedimentary rocks.

The geologists believed formerly that the Earth gradually and continuously has grown cooler. This theory, however, struggled with the difficulty that certain cold time intervals, ice periods, were succeeded by warmer epochs. To begin with efforts were made to surmount this obstacle by assuming that an ice period on the north hemisphere was counterbalanced by a warm period on the south hemisphere and vice versa. In this manner, the mean temperature for the entire surface of the globe might possibly have been continuously decreasing although fluctuations had occurred on the two hemispheres. But this view has proved untenable, because the ice period has left traces also within the tropics, near the equator, as on Kilimanjaro, in New Guinea, and so on. It is now practically agreed that the last great ice period was characterized by a temperature between 4° to 5° C. (7° to 9° F.) below the present all over the surface of the Earth. This determination has been accomplished by measuring the difference in height between the terminals of the glaciers at present and the lowest points where their grinding action has left obvious traces. The ice coverings of North Europe, North-East America, South America, along the coast of Chile and in Argentina, as well as on the southern island of New Zealand, all appear to have existed simultaneously. Also during earlier eras, for instance during the Algonkian and the Permian epochs, ice periods have occurred. The latter, which was felt in Australia, India, and South Africa, is called the Gondwana-time. This period was formerly supposed not to have caused any temperature drop except in the tracts mentioned. Later investigations lead us to believe, as asserted by Holland in his presidential address to the geological section of the British Association at its 1914 meeting, that also this ice period simultaneously embraced the entire globe.

As the Algonkian time belongs to the oldest epochs of geological history it appears that the temperature on the Earth as long as life has existed on our planet on the whole has been nearly constant, with important alternations, however, of warm and cold periods. For an explanation of these fluctuations, our well-nigh only recourse is the assumption that the heat-conserving quality of the atmosphere has changed by virtue of a varying composition. Warm periods occurred when carbon dioxide was abundant in the atmosphere due to volcanic activity, cold periods again accompanied a paucity of carbon dioxide. With rising temperature, the percentage of water vapour in the air also increased, affording further protection against radiation loss of heat.

Thus, it would seem as if the mean temperature of the Earth's surface hardly had changed to any extent worth mentioning during immense spans of time estimated to about 500 million years. Nevertheless, a slow process of cooling proceeding toward the centre of the planet probably takes place. Ever growing quantities of matter are transported from the interior of the earth through volcanic action. Sedimentary deposits increase continuously while the interior becomes hollow. As a result the crust must gradually settle, causing large cracks in the process. For these weakened places the volcanic products show a special liking and the craters are strung out in lines along such fissures. In other places, where volcanic action is less pronounced, hot springs appear instead, generally emitting carbon dioxide in abundance, occasionally also sulphurous acid and sulphuretted hydrogen. The dislocations in the crust also take place along these cracks accompanied by earthquakes. The study of these various phenomena has enabled us to map out the fissures, which generally radiate in nearly straight lines from one point, the so-called centre of collapse, as the cracks in a pane of glass issue from the point of breakage caused by a swift blow. We shall later see that such breakage lines and centres of collapse are common on all stellar bodies which possess a solid crust and are observable from the Earth.

We may now easily form an idea of the general trend in the development of the atmosphere. The gases originally present were all, except the hydrogen, the nitrogen, and the rare gases, strongly absorbent of light and in particular of heat. It is, therefore, natural that the planets which have not formed a solid crust possess a strongly absorbing vapour-shell, as indeed is the case with the large planets (compare Figure 13). The crust once formed and the air gradually purged of these gases, thanks to the sunlight, so that mainly nitrogen and oxygen, small quantities of the rare gases, and carbon dioxide besides water remained, the temperature fell rather rapidly. Carbon dioxide formed the last effective heat-conserving ingredient. As the crust grew thicker, the supply of this gas diminished and was further used up in the processes of disintegration. As a consequence the temperature slowly decreased, although decided fluctuations occurred with the changing volcanic activity during different periods. Supply and consumption of carbon dioxide fairly balanced as disintegration ran parallel with the proportion of this gas in the air. But evolution on the whole can only proceed in one direction toward a final cooling of the Earth. This must occur if for no other reason because the store of energy in the Sun and therefore its radiation must slowly decrease. With deepening crust and disappearing carbon dioxide vegetation must ebb, and with it the production of oxygen. This gas also partakes in the general disintegration through oxidation of iron protoxides in the mineral rocks. The oxygen portion of the air must therefore finally reach its maximum and start on the decline. Calculations point to the conclusion that the carbon dioxide of the air would be consumed in a few tens of thousands of years if new supplies were not furnished from the interior. Water is also absorbed in the processes of decay as hydrated compounds are formed, increasing in quantity with falling temperature. As the amount of water in the ocean is immensely larger (about 50,000 times) than the stores of carbon dioxide in the air and in the seas the lack of the latter will undoubtedly first become serious. But a slow desiccation of the planet must subsequently take place, and will proceed at an accelerated rate with the continued cooling of the Earth. Then the vapour in the air and consequently precipitation will wane. Then, as during the ice periods, mighty ice caps will cover the poles and impound a large portion of the water in the ocean. Finally, the entire planet, perhaps after having harboured life during trillions of years, becomes an ice waste with a few cracks in its hard crust through which warm and acid vapours rise and create small melted areas characterized by a darker colour than the desert and ice-landscape in general. Organic life lacks the conditions for existence and ceases therefore to cheer the planet with its interesting variations. The planet is dead but continues in obedience to gravitation to describe its orbit in space.

Chapter 6

THE PLANET MARS

Through the works of Schiaparelli, Flammarion, and Lowell the vivid interest of the general public has been directed toward our neighbour planet Mars. Several investigators, Flammarion and Lowell among them, assert with full confidence that Mars has intelligent inhabitants, who have built and maintain the curious "canals," which, it is stated, could not have been created except by intelligent beings far superior to man.

Air, water, and sunshine exist there, says Flammarion in his well-known great work, *La Planète Mars* (1902, page 515). "It appears incongruous to us to condemn a world like Mars, where all the conditions for life exist, to such a fate" (to be a dry desert). No doubt sentiment and the desired result play a part in all such deductions, as indeed the words chosen by Flammarion would indicate.

As contrasted to the Earth, Mars is, on the other hand, considerably further removed from the Sun, whose radiation therefore on Mars possesses only 43 percent of its intensity on Earth. Judging by this fact, the mean temperature of Mars should fall far below that of the Earth and considerably below the freezing point of water and under such conditions it is hard to imagine a vegetation near the poles of Mars as Lowell does in his volume, *Mars as the Abode of Life* (1909), or even in the neighbourhood of the canals anywhere on the surface of the planet as assumed by Flammarion.

With such ideas in vogue we can well understand that the astronomers would point their lately extremely sharpened instruments toward our ruby-coloured neighbour in the sky when, in 1909, it passed very close to the earth under conditions particularly favourable to accurate observations, more so in fact than they had been during the seventeen preceding years.

Numerous astro-physicists, among them the world's foremost representatives of their discipline, have repeatedly turned their spectroscopes toward Mars in order to ascertain whether water vapour was present there or not. In the spectrum of the Sun we find several so-called "rain-bands" due to the fact that the light before it reaches the apparatus has passed through the moisture of the air. The more humid the air the more strongly developed these rain-bands are. If we direct the instrument first on the Moon, which lacks an atmosphere and therefore also moisture, and then on Mars, which for the sake of simplicity we assume standing close to the Moon-disk, a difference should appear in the spectra of these two bodies, provided moisture is present in the atmosphere of Mars. The rain-bands ought to be more pronounced in sunlight that has passed the atmosphere of Mars (passed it twice as the light is reflected by the surface of the planet) than in light reflected from the naked Moon. The bands appear of course in both spectra as the light on its final stage to the spectroscope passes the atmosphere of the Earth, which never is free from moisture. In this manner Huggins and Janssen, scientists of world-wide fame, believed that they had demonstrated the presence of water vapour on Mars. Campbell, on the other hand, the prominent director of the Lick Observatory, made similar investigations of the planet in 1894, and so did a French astronomer, Marchand, in 1896 and 1898, both under unusually favourable circumstances as the former installed his instrument 1283 m. (5200 ft.) and the latter 2860 m. (9370 ft.) above sea level, but neither found any indication of moisture in the atmosphere of Mars.

It is evident that the observations would be far more accurate if we could eliminate the moisture from the atmosphere of the Earth, in which case no rain-bands would appear in the spectrum of the Moon. It would then be unnecessary to compare the two spectra; we would only have to determine whether rain-bands were present in the spectrum of Mars or not. We can never entirely avoid the water vapour of the air, but its influence may be greatly reduced by making our observations from high mountains or in desert climates where the air is comparatively dry, that is free from water vapour. Investigations undertaken where the air is dry deserve therefore more confidence than those handicapped by greater humidity. The observations by Campbell and Marchand fall in the former category, and it would, therefore, appear that the presence of water vapour on Mars to any extent worth mentioning is highly doubtful.

In later trials, Campbell and Keeler have employed an improved method, using photographs of the spectra on sensitive plates, but neither has succeeded in discovering any water vapour in the atmosphere of Mars.

Obviously photography offers a great advantage over direct ocular observation. The two pictures may be placed side by side and very accurate measurements may be made at leisure. We may also choose the moments for exposure when the two stellar bodies stand equally high over the horizon so that the sunlight reflected from them traverses equal distances in the humid atmosphere of the earth.

It now devolved upon Lowell to test his theories by means of the magnificent resources at his disposal in Flagstaff observatory in the desert of Arizona 2200 m. (7200 ft.) above sea level. In the months of January and February the dew-point there is about -7° C. (+19.4° F.) *i. e.*, each cubic meter (1.3 cu. yds.) of air contains 2.8 grammes (43.25 grains) of water vapour while saturated air at zero temperature (32° F.) holds nearly twice this amount or 4.8 grammes (74 grains) per cubic meter (1.3 cu. yds.). Slipher, working in this observatory, pushed the sensitiveness of his plates to the utmost then obtainable, and photographed the spectrum of Mars in January and February, 1908. He found that the most important rain-band always was more prominent in the spectrum of Mars than in the spectrum of the Moon photographed later during the same night. Peculiarly enough, it was only the rain-band designated "A," and located in the red spectral field, that was of a marked difference in the two spectra. Other bands gave no indication of the presence of water vapour on Mars. This result did not directly contradict the conclusions reached by Campbell and Keeler, also by means of photography; they had investigated other bands than “A.” The “A”-line might therefore possibly be more sensitive to water vapour than the others.

Slipher’s discovery was considered so valuable that it must be employed to the limit. The well-known physicist Very was therefore called in consultation; he made careful measurements of the intensity of the “A”-lines on the various plates and calculated that the atmosphere of Mars contained 1.75 times as much water vapour as that of the Earth at the point of observation. If we desire to determine the proportion of water vapour in the air at the surface of Mars from this statement we may figure in the following manner. The amount of water vapour in a vertical column of air one square meter (1.2 sq. yds.) in section is, according to Hann, 2500 times the amount in a cubic meter (1.3 cu. yds.) at the surface of the earth. At the time of the observation, the latter amount was 2.29 grammes (35.4 grains); on each square meter (1.2 sq. yds.) of the ground rested therefore 5725 grammes (12 lbs. 11 oz.) of water vapour. That the quantity of water is not larger, although the depth of the atmosphere far exceeds 2500 m. (1.5 miles), is due to the fact that the temperature rapidly decreases with distance from the ground. On Mars, the temperature ought not to fall so quickly with change in height because the intensity of gravity there is 2.68 times smaller than on the Earth. The temperature drops there 2.68 times slower with ascent in the atmosphere, and a column of air on Mars one square meter (1.2 sq. yds.) in section should therefore contain 6680 times as much water vapour as a cubic meter (1.3 cu. yds.) at its surface. As Mars did not stand in zenith, the distance traversed by the light-ray in the atmosphere was greater—in fact 1.43 times greater than if such had been the case. A column of air in the direction of the light-ray, one square meter (1.2 sq. yds.) in section contained therefore 8 grammes (18 lbs. 3 oz.) water vapour. In the atmosphere of Mars which the light passed in a vertical direction there was, if we are to believe Very, 1.75 times as much, or 14,300 grammes (31 lbs. 8 oz.) and in a cubic meter (1.3 cu. yds.) at the surface of the planet, consequently 6680 times less, or 2.14 grammes (33.1 grains). The corresponding dew-point is then, according to this determination, -10.3° C. (+13.5° F.). It is agreed upon that a desert climate prevails on Mars. It might at the time of the observation conform to the extremely dry climate at Salt Lake City in the height of summer when the humidity there is only 31 percent of saturation. Under such conditions saturated air at noon in the equatorial belt on Mars should contain 7 grammes (108 grains) per cubic meter (1.3 cu. yds.) corresponding to a temperature of 5.3° C. (41.5° F.).

It must be admitted that this was not very encouraging to Lowell. If the temperature in the middle of the day, when the sunlight falls perpendicularly on the surface of the planet, rises only to about 5° C. (41° F.), the mean temperature for twenty-four hours, even in the midst of the summer, must in this entirely clear, light air be far below freezing and vegetation on Mars is therefore not very well conceivable. In spite of this, Lowell saw in Slipher's measurements a confirmation of his theory that Mars is the abode of an intelligent race that utilizes, in their wrestle with existence, a verdant vegetation pushed even into the polar regions.

Campbell, however, went one step further than Slipher. In August and September, 1909, Mars occupied a position in the sky particularly favourable to observations. Campbell decided to benefit thereby. With the support of a rich patron of science, a Mr. Crocker, who on several occasions has made magnificent contributions toward astronomical research, Campbell equipped an expedition to Mount Whitney in California, 4425 m. (14,502 ft.) high and the loftiest peak in the United States. He was accompanied by an able scientific staff, the most prominent of which were Dr. Abbot, head of the observatory belonging to the Smithsonian Institution, and a well-known German astronomer, Albrecht. The members of the expedition were affected by mountain sickness and suffered many severe hardships when the wind was high, reaching about 25 m. per sec. (56 miles per hour), and at the same time cold, falling below zero (freezing) during the night. The barometric pressure was only 447 mm. (17.6 inches). During the nights, when the observations were made, the water content of the air fell to between 0.5 and 0.9 grammes (7.7 grains to 13.9 grains) per cubic meter (1.3 cu. yd.) or 2.5 to 4 times less than Slipher had to contend with. The spectra of the Moon and of Mars were photographed in close succession, two exposures being made in each case. The band "A" was plainly visible on several plates. No indication of greater prominence of this band in the spectrum of Mars could be found. Other rain-bands were also investigated with the same result. Neither were the characteristic bands of oxygen stronger in the spectrum of Mars, than in that of the Moon. Slipher believed that he had discerned a difference, although of a hair's breadth, which would indicate the presence of oxygen in the atmosphere of Mars. The conclusion itself is not improbable, but the amount of oxygen there is in any case considerably smaller than in the Earth's atmosphere.

Several statements by Campbell, as well as Slipher's observations, indicate that a difference ought to have appeared between the spectrum of Mars and that of the Moon if the water content in Mars' atmosphere had been the same as in the Earth's at the time of the observation. This content, as stated before, was about 3 times smaller on Mount Whitney than at Flagstaff. At the latter place, the measurements gave 1.75 as the ratio of water vapour on Mars to that on the Earth. The amount of water vapour with the Sun in zenith on Mars should therefore, according to Campbell's observations, only reach 0.4 gramme (6.1 grains) per cubic meter (1.3 cu. yds.) corresponding to a dew-point of -28° C. (-18.4° F.) or to an actual temperature of -17° C. (+1.4° F.) allowing also for a desert climate with only 31 percent saturation. This temperature is probably higher than the mean for a summer day as the observations were made at noon on Mars.

It must now be evident that we should consider Mars as unfit to harbour living beings. There is possibly a slight amount of oxygen in the thin air but the extremely low temperature and the scant supply of water vapour form insurmountable obstacles to the subsistence of even the simplest forms of life in the equatorial regions on Mars. The temperature difference between day and night must be enormous on account of the desert climate. Even if life could develop during the day, which has nearly the same duration as with us—Lowell fixed it at 24 hours, 37 minutes, 22.6 seconds—and during which the temperature possibly might rise above the freezing point, it would nevertheless be destroyed without mercy by the bitter frost at night.

Campbell has offered an explanation of the indications of water vapour on Mars, apparent on Slipher's photographs. An analysis of the latter's observations shows that the Moon was photographed about four hours later in the night than was Mars. On all occasions except one, clouds appeared in the sky. This indicates the presence of moisture in the air, so that the humidity should change with the temperature which latter rapidly falls in the course of the night. Campbell, himself, found during the clear nights, when he made his observations, that the humidity in the hours of the night up to midnight falls to a fraction—a half or a third—of its original value an hour or so after sunset. This rapid temperature drop is probably confined to the strata immediately above the observation point but the moisture is strongly concentrated downward so that this change in humidity undoubtedly should have been taken into account. Or better, observations should be avoided in the beginning of the night and the Moon and Mars photographed as soon after each other as possible, precautions taken by Campbell but not by Slipher. That the latter found less traces of water on the lunar photos than on the Martian ones is, therefore, probably due to the fact that the former were taken about midnight but the latter not long after sunset, when the atmosphere contained much more water vapour. Thus, we learn how a small slip, more obvious to the meteorologist than to the astronomer, may spoil a labour otherwise done with extraordinary care.

To Campbell's critique Very answered by the suggestion that the meteorological conditions during the Mount Whitney observations should have been exceptionally unfavourable. The entire south-west of the United States and the north of Mexico were visited at that time by cloudy weather and heavy downpours. Very contends that this humidity should partly have extended to the high strata above Mount Whitney and therefore rendered the calculation of the moisture content of the air entirely unreliable.

Simultaneously (August, 1910) new measurements were published of Slipher's photo-plates from February, 1908, which Very had examined. The result was now that the rain-band "A" was 2.5 times more pronounced in the spectrum of Mars than in that of the Moon. Furthermore, the oxygen absorption-band "B" was 1.5 times stronger for Mars than for the Moon.

Great quantities of water vapour and oxygen should, therefore, undoubtedly exist in the atmosphere of Mars.

In the meantime Campbell had not been idle. The difficulty with the older measurements consisted in the fact that the absorption line of water vapour in the atmosphere of Mars occupies the identical place of the line due to vapour in the Earth's atmosphere. There exists, however, a method, as already pointed out by Campbell in 1896, of separating the two, which method is available when Mars either approaches or departs from the Earth with sufficient velocity. The latter could be determined both from the known motions of the two planets and from the displacement of certain spectral lines of the Sun. These two determinations were in almost perfect agreement; for instance January 26–27, 1910, astronomical calculations gave a relative velocity of 19.1 km. (11.86 miles) per second and spectroscopical measurements 19.2 km. (11.93 miles) per second, while on February 3–4 the relative velocity was 18.1 km. (11.24 miles) a difference of 1 km. (.62 miles) per second. This trial shows the accuracy of the method. Among the absorption lines of water vapour and of oxygen there was, however, none due to the atmosphere of Mars. Campbell assumes that such lines would certainly have been visible if they had been only one-fifth as strong as the so-called tellurian lines. The advantage of this method is evidently that the "Martian" and the "tellurian" lines lie close beside each other on the same plate so that differences in sensitiveness, exposure, and atmospheric conditions are entirely eliminated.

From these and the following data we may calculate water content and temperature of the atmosphere on Mars anew: water vapour at the point of observation was 1.9 grammes (29.3 grains) per cubic meter (1.3 cu. yds.), zenith-distance of Mars 55° and incident as well as reflected sun-rays formed an angle of 70° with the surface of Mars; hence, the amount of moisture at the surface was only 0.12 gramme (1.85 grains) per cubic meter (1.3 cu. yds.), corresponding to -38° C. (-36.4° F.) for saturated air and to -27° C. (-16.6° F.) for air of 31 percent saturation. Oxygen content per cubic meter (1.3 cu. yds.) at the surface of Mars would be only a sixteenth part of corresponding numerical value on the Earth. This determination is more accurate than any of the previous ones and reduces the temperature another 10° C. (18° F.) below the lowest value derived earlier in this chapter. We should remember, however, that, during the trial of September, 1909, the sun stood practically in zenith on Mars, while in January and February, 1910, we are concerned with a point where sunrise had occurred about four and a half hours previously. The latter observation should give a value close to, but slightly above, the mean diurnal temperature on Mars.

No determination comparable in precision with this one by Campbell appears to have been made. We must therefore recognize it as conclusive.

We may easily calculate the surface temperature of a planet from the intensity of the solar radiation received, or insolation, provided the surrounding vapour shell contains no heat retarding gas. The most important gases of this kind are water vapour, which, as we just have seen, is very sparse in the atmosphere of Mars, and carbon dioxide, of which there probably also, for reasons stated below, is only a scant supply in the martian gas shell. Such calculations were first performed by Christiansen of Copenhagen, who assumed 2.5 calories as the solar constant on Earth, *i. e.*, the amount of energy received through insolation per minute by each square centimeter (.15 sq. in.) of the Earth's surface when at right angle to the radiation and on mean distance from the Sun. On Mars, the radiating energy received under similar conditions is only about 1.1 calories. The surface of the planet is heated until it radiates as much energy into space as it receives from the Sun. In this way we obtain an average temperature of -37° C. (-34.6° F.) for the entire surface of Mars. The regions, exposed to the Sun in zenith at noon, might, if heat were not conveyed therefrom, possibly reach a daily mean temperature of +8° C. (46.4° F.) and perhaps slightly more at noon. Probably not even the freezing point is reached, as the heat is rapidly carried away by the freely circulating air. The above-mentioned mean temperature of -37° C. (-34.6° F.) seems on the whole to agree well with the observations by Campbell on Mount Whitney.

Recent accurate determinations of the intensity of the solar radiation by Abbot, K. Ångström and others, indicate that it has been estimated about 20 percent too high. If we take the solar constant to an even 2.0 calories, which is a trifle high, we reach the conclusion that the mean temperature on Mars would fall about 50 degrees below freezing. Equatorial regions might then reach an average of -8° C. (+17.6° F.) and at noon the temperature might possibly rise slightly above zero (32° F.). A higher temperature yet might be attained at the pole where the Sun during the summer remains for months above the horizon or a high mark of +8° C. (46.4° F.) provided no heat were carried away by air currents. Such losses naturally must occur, and the temperature probably hovers around freezing. At the Martian poles we might possibly imagine the existence of some low forms of vegetation (snow-algæ, etc.) during the height of the short summer.

Figure 15. Narrow bay on the left side of Segelskar, east of Hango in Finland. The bay owes its existence to the fact that the ice has laid bare a territory coursed through by fissures. Photo by I. I. Sederholm.

When we hitherto on the authority of Lowell, Very, and others, have assumed an average temperature of +10° C. (50° F.) on Mars, we have done so on the supposition that the atmosphere of the planet contained great quantities of heat-conserving gases. This assumption appears to be no more tenable than the belief in the high temperature on Mars. After all, the temperature is probably about 10° C. (18° F.) higher than our last calculation would indicate—or about -40° C. (-40° F.)—because the air on Mars is very clear and admits, therefore, all sun-rays, retaining also a fraction by virtue of what little water vapour, carbon dioxide, and other heat-conserving gases there may be present in the atmosphere. The mean summer temperature at the martian equator (-27° C. or -16.6° F. acc. to Campbell's data) would then lie about 13° C. (23.4° F.) above the mean for the planet. This agrees closely with conditions on the Earth where the highest mean in July at the equator is 27° C. (80.6° F.) and the mean for the earth 16° C. (60.8° F.).

We are consequently obliged to revise in their entirety our ideas about Mars. The belief that organic life (green vegetation) causes the colour of the so-called seas on Mars, as assumed by Lowell, or that the red tints belong to the gorgeous dress in which autumn arrays the plants before their leaves are shed under the attacks of frost, as intimated by Flammarion, must nowadays take its place in the shadowy realm of dreams.

Those who do not believe that the so-called canals are real waterways, devoted to freight carrying and irrigation, or illusions, which conception the photographs contradict (for example Figure 18), generally consider that they signify cracks or fissures. As in the crust of the Earth, they generally run in nearly straight lines or in regularly bent curves (Fig. 17 and 17a). Flammarion mentions that the renowned physicist Fizeau looked upon the "canals" as cracks in the ice-coverings of the oceans on Mars. Penard, in 1888, expressed the more likely opinion that they correspond to the fissures in the crust of the Earth. Flammarion contends that such fissures do not have the rectilinear configuration of the "canals." This is completely in error, as shown on the map here reproduced (Fig. 16). It is also stated that they are so inexplicably long, for instance the canal Phison is 2250 English miles (Lowell) or 3620 km. in length. The longest known earthquake crack along the entire length of which a dislocation has taken place *at one time* is 600 km. (373 miles) in extension; the violent earthquake in California, 1906, originated from this crack. There is no doubt, moreover, that a great fissure in the Earth follows the coast of Chile from Arica to the Strait of Magellan in a nearly north and south direction for a distance of about 32 parallels or 3560 km. (2210 miles). This fissure is almost as long as Phison on Mars. Such cracks exist along the entire coast of the Pacific Ocean. As yet, we do not know their position in much detail, because long stretches run below the sea or through territories not yet occupied by civilized people. As an example of a small fissure, a picture taken by Sederholm from Segelskär, east of Hangö in the Baltic, may serve (see Figure 15). As the studies of earthquakes are prosecuted with increasing interest in later years, fissures of all dimensions will undoubtedly soon be discovered. The solid crust on Mars is, furthermore, somewhat thicker than that of the Earth as the cooling of that planet has progressed further. The sections broken off at the bursting of the Martian crust ought therefore to be much larger both in breadth and in length. No doubt, the facts that the intensity of gravity on Mars is only three-eighths of its intensity on the Earth and that the curvature of the Martian surface is twice as sharp as the Earth's contribute to this result. Imagine two vaults, one built with higher and broader wedge-shaped stones than the other and with half the radius and furthermore loaded only one-third as heavily as the other, and it will become evident that we can permit a much larger span in the former than in the latter case without fear of collapse. In other words, it requires a much more extensive caving or shrinkage of the molten mass beneath the crust of Mars to cause a rupture than under the terrestrial crust.

As a consequence, the fissures on Mars ought to be longer than the corresponding formations on the Earth. A thorough study of the large fissure in Calabria shows that it consists of a veritable network of smaller straight cracks (as is apparent on Figure 16, which is taken from a work by the well-known Finnish geologist Sederholm). On this map the radial cracks (see *Worlds in the Making*, Figure 16), charted by Suess, are also shown, and their direction under the sea designated by dotted lines. The sketch in the upper left corner of Figure 16 is in striking similarity to a picture drawn by Schiaparelli in Mercator's projection of the planet Mars (see map at the end of book). We notice on both, the numerous equidistant lines corresponding to parallel cracks and duplex canals. Not every fissure has its parallel and not every canal its mate—generally only one of the latter is visible and sometimes both disappear.

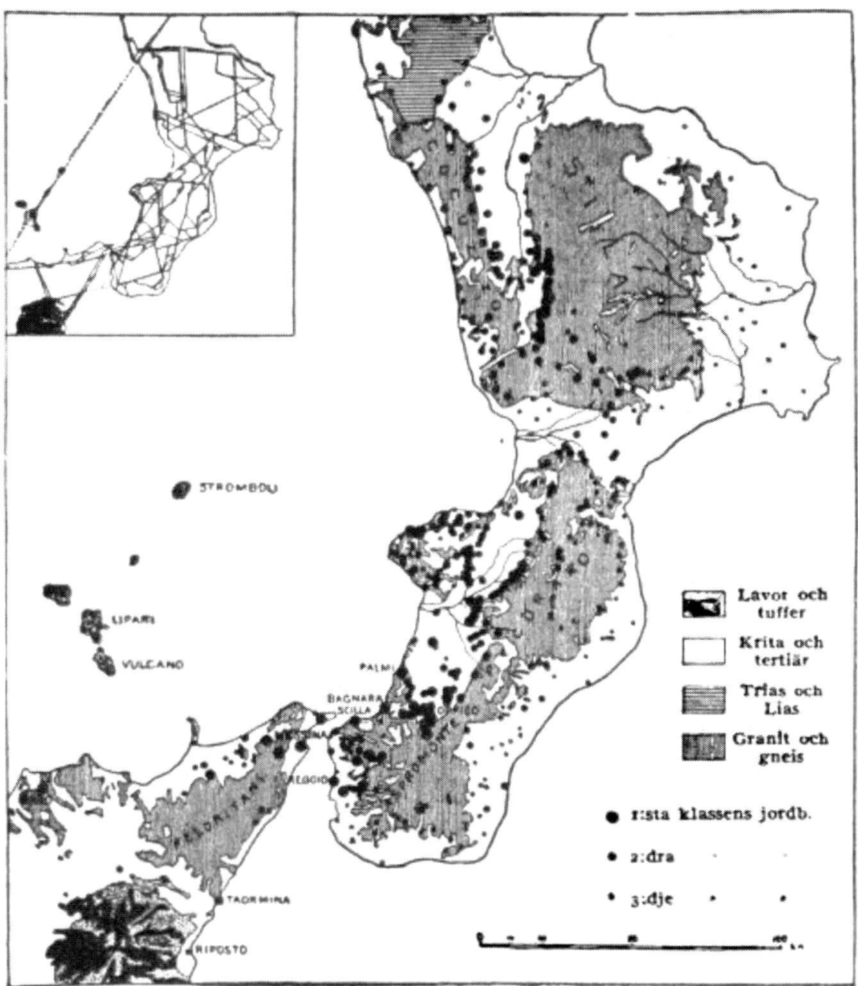

Figure 16. Earthquake centres in Calabria and in Sicily. On the larger map the damaged localities are indicated, on the smaller the most prominent tremor lines. Drawn by I. I. Sederholm.

As the radial cracks in the drawing by Suess if extended meet in the Lipari Islands, so also several canals on Mars run together in a so-called lake (Lowell called them "groves" or "oases") which evidently is a centre of collapse (many appear on Figure 17). It is plain that all crossings of the "canals" are not necessarily such centres of collapse. (See maps Figures 17 and 17a at end of book).

We shall consequently assume that the canals on Mars correspond to the geological dislocation fissures on the Earth. Along these fissures emerge the gases liberated in the cooling process on both planets; which are similar gases to those which escape through the volcanoes. These vapours are primarily water, next carbon dioxide and, in considerably smaller quantities, sulphuretted gases and hydrochloric acid. They discharge through cracks in regions which, geologically speaking, not so long ago were the scene of volcanic activity. In the dislocation-grooves, lakes, and water courses are often formed, as we may observe in several places in Sweden, for instance near Stockholm.

Assume now a gradual cooling of our earth. Most territories are covered by stratified, comparatively light rocks. To the dislocation fissures water gathers from the surrounding strata and occasionally from the interior, partly washes away the loose material and transforms the fissures into furrows, generally with flat bottoms. Dissolved salts are carried to the sea. As cooling proceeds, the ocean commences to freeze. Each summer the surface melts to a certain extent, as is the case now in our polar regions. Finally, the entire ocean freezes to the bottom, the ice is now to be considered as a kind of rock, flexures and dislocations cease and the ice assumes a smooth surface. In the strong sunlight during the summer this surface thaws, as do the water-courses on the mainland, and these continue to carry their salts to the open surface water. At the approach of winter the latter solidifies again but not as the water in our inland lakes from the top but from the bottom, as ordinary sea-water possesses its greatest density below the freezing point while the opposite is true of fresh water. The consequence is that the ice foundation grows upward and as the surface water becomes increasingly shallow it is turned into a concentrated salt solution. With a further drop in temperature the ice formation is accompanied by crystallization of the salts.

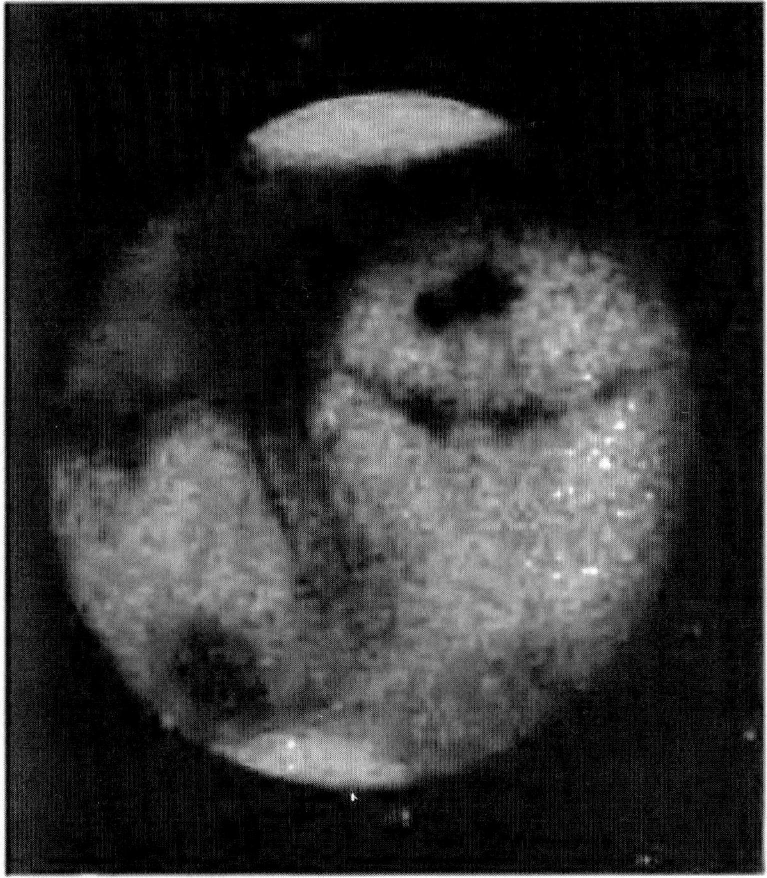

Figure 18. Photograph of Mars, enlarged and retouched; taken by Lampland.

Something similar takes place on the Martian mainland in its flat river basins, which correspond to the salt lakes in our deserts. On account of the bitter cold and the consumption of the water in the process of disintegration (the carbon dioxide has been largely used up in the same manner), precipitation has almost ceased on Mars and most of the water in circulation emerges from the interior of the planet along the fissures. As it contains hydrochloric acid and carbon dioxide it extracts from the soil salts, such as the chlorides of sodium (common salt), of calcium and of magnesium, all present in common sea-water to which it was brought by the rivers. The compounds of calcium and magnesium are not precipitated as carbonated salts through the medium of crustacea as is the case on earth. The strong solar radiation during the summer partly evaporates the water into the thin air, leaving the salts behind. On account of the low temperature, this vaporization on Mars is probably slower than on the Earth. Along the cracks in the crust, a kind of dry salt-lakes are formed similar to the generally shallow and occasionally dry lakes common in the deserts of Central Asia as described by Hedin. We know that Mars possesses a pronounced desert climate. There finally remains in the lowest sections of the water courses a concentrated salt solution, which parts with its water more and more reluctantly, so that the salts which most strongly hold the water crystallize at the deepest points. If the winter's cold is sufficiently severe (below -55° C. or -67° F.) ice is extracted even from the most concentrated solutions, which mainly contain chloride of calcium. In spite of such extreme temperatures, evaporation into the rare atmosphere is not negligible and the ice crystals partly vanish, to reappear in the coldest regions of the planet, that is, around the pole which at the time is turned away from the sun. On the ocean, now frozen solid throughout, a polar-cap of snow and hoar-frost is formed which finally reaches as far as the 38th parallel on the southern hemisphere (see Figures 18 and 19), where winter occurs when Mars is most removed from the sun, and to the 58th parallel (see Figure 19) on the northern hemisphere where winter reigns while Mars is nearest to the Sun and consequently not quite so cold. Similar conditions obtain on the Earth although not to such a marked degree.

In the vicinity of the snow-white polar cap, whether there be continent or sea, bodies of water occur with solidly frozen surface covered by crystals of very hygroscopic salts, such as the chlorides of calcium, magnesium, and sodium. When the summer warmth returns and the polar cap is heated, the hoar-frost evaporates, and the now comparatively humid air spreads over the surrounding territory. We observe also frequent mist-formations in these places. The ground near the edge of the polar snow assumes then often a dark hue on account of the moisture (Figure 19).

Occasionally canals and lakes appear in the polar cap (see Figure 20). This is evidently due to hot emanations along the cracks. The moist air sweeps over the salts, which then absorb water and dissolve into concentrated solutions. New quantities of water vapour are supplied from the pole as they distill over toward the other pole, where winter now exists, and push on toward the equator which they finally pass. In their course they dissolve the salts in the depressions along the fissures and particularly at the deep crossings where the centres of collapse or the so-called "oases" are located. Lowell has observed that the "canals" in this manner gradually "liquify" from 78° N. Lat. to the equator in fifty-two days.

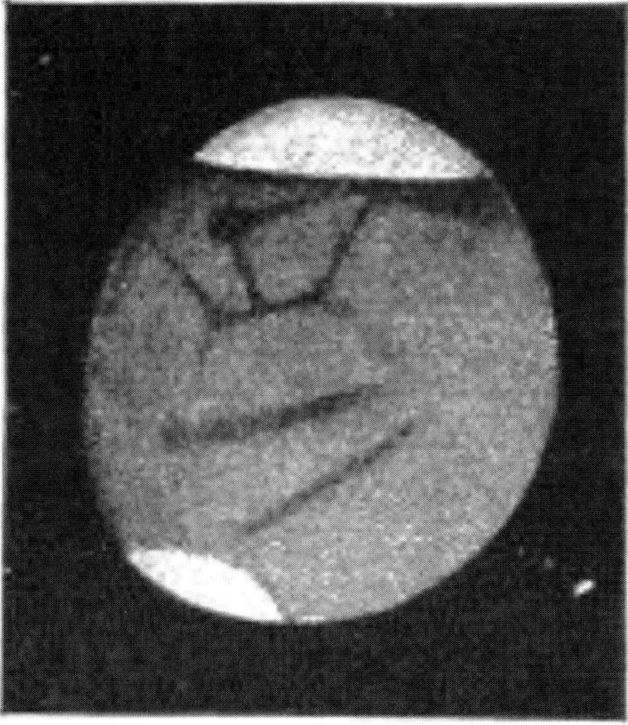

Figure 19. Appearance of Mars April 8, 1907, as observed by Quénisset. Below one pole a dark line is plainly visible.

Figure 20. The South Pole spot on Mars observed by Jarry-Desloges July 10, 1909.

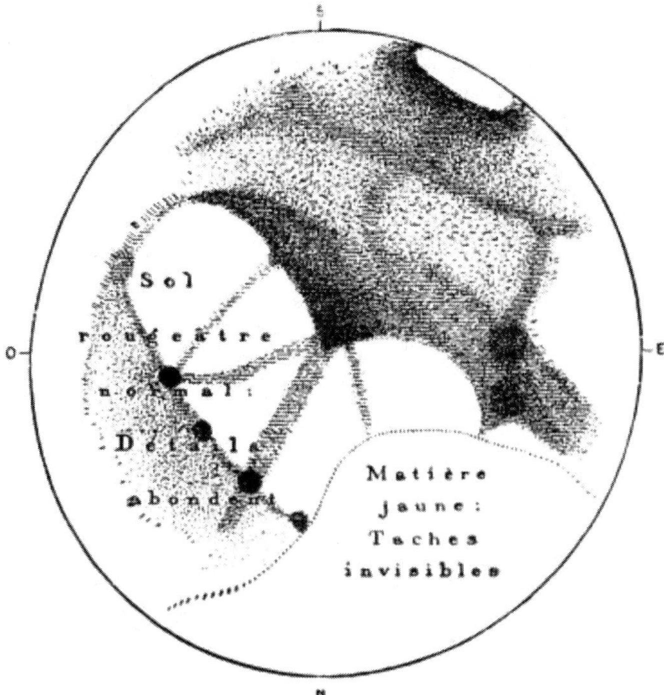

Figure 21. The appearance of Mars during an observation by Antoniadi in 1909. The whole disk is somewhat hazy; below, the details are hidden by clouds of sand.

The canal theory presents great difficulties to the explanation of this curious phenomenon. In order to make the water flow it must be assumed that the surface of Mars is entirely smooth or at least very nearly so and that the inhabitants convey the water melted at the poles through pumping stations. The canals vary in width; according to Lowell, their mean is 16 km. (10 miles), according to Flammarion between 300 and 60 km. (185 and 37 miles) which latter estimate probably is too high. The same canal differs widely in breadth in successive years and sometime disappears altogether. When the supply of water vapour is scant, only the most hygroscopic salts are dissolved, *i.e.*, those deposited in the deepest furrow of the canal, but when the moisture sweeping over the canal is more abundant the broader portions absorb water, darken and thus become visible. The same holds true in regard to the inland lakes ("oases"). As the water vapour diffuses in the air, the canal becomes liquid along its entire length independent of the altitude of its various parts.

All agree upon the desert climate of the mainland on Mars. Like most deserts on the Earth, it is, therefore, probably a table-land, where one plateau mounts above the other, each one nearly level. By the action of the wind, the upper layers have been transformed into fine sand. On the dead planet no further sediments are deposited by the sea. The only accretions to the planet are meteorites and cosmic dust which slowly rains down. It contains among other substances iron, partly metallic and partly in the form of protoxides (which have a light green colour).[5] The oxygen in the atmosphere of Mars transforms these compounds to ferro-oxide which has different colours according to its coarseness, but generally is ochre. The surface of Mars is also described as possessing this colour. Dross has, therefore, assumed that the Martian soil is mingled with ferro-oxide. The finest dust, however, is yellow while larger crystals tend toward violet. We often observe on Mars that the details are covered by a yellow veil. This is of course finely powdered ferro-oxide probably mixed with less coloured sand which the desert wind whirls up over large portions of Mars. Vast

---

[5] In the deepest portions of the oceans on the Earth where no sediments from the coasts are deposited, large accumulations containing iron compounds have been found and the presence of certain minerals therein points to their meteoric origin.

sections of the planet bore such wrappings in the autumn of 1909, as observed and described by Antoniadi in Paris (see Figure 21). Similar observations have previously been made by W. H. Pickering and others.

As a rule, only the central and the polar regions of the surface of Mars can be seen. Territories near the equator more than 40 to 50 degrees removed from the point in line with the Sun and the centre of the planet are generally hidden behind a thin, white veil of mist. As soon as the Sun leaves the zenith and reaches half-way to the horizon the moisture of the air is precipitated near the ground. This shows that the planet does not possess any quantities of heat-conserving vapours in its gas shell. The mist does not extend to the poles, whose white caps always appear distinctly, because the Sun cannot greatly affect the evaporation in regions where the Sun's altitude is neither very high nor very variable. The same holds true for other snow-covered tracts, even if they are not located in the immediate vicinity of the pole.

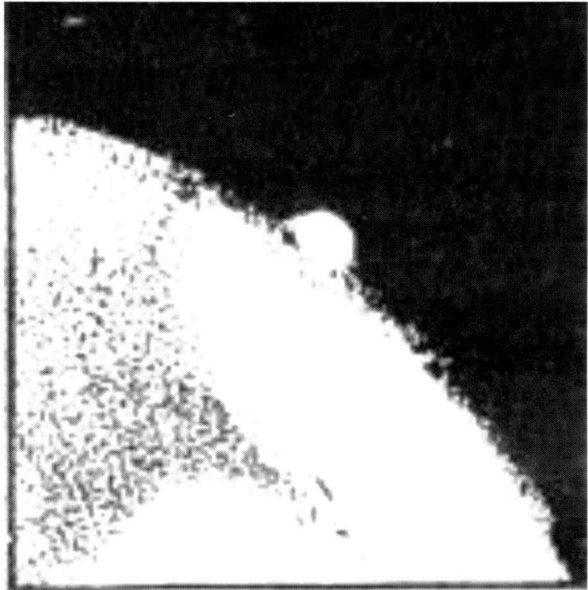

Figure 22. Cloud at the upper right edge of Mars observed by Molesworth, March 7, 1901.

When the supply of water vapour is scant, only the most salient canals come into view. As a rule they do not then appear double, as one of the mates is always less prominent. Lowell showed, he believed, that it always is the same canal out of a pair which first comes to sight and that its position always remains unchanged in contrast to Schiaparelli who has reached the opposite conclusion. This, of course, is quite natural.

On account of the small amount of water vapour in the atmosphere of Mars true clouds are rare. Figure 22 shows such a cloud at the edge of the planet. The aforementioned mists are often called clouds, for instance by Pickering.

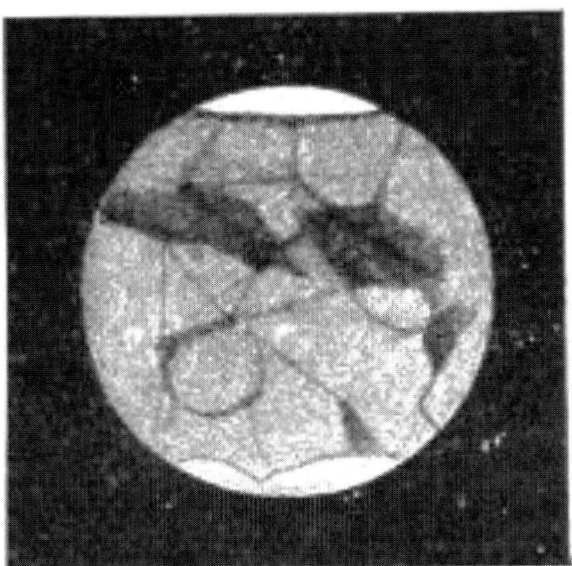

Figure 23. Mars as observed by Lowell, July 11, 1907. Even the dark portions appear coursed through by "canals."

That elevations really may be found on Mars is evident from the fact that snow or hoar-frost often remains in patches near the pole and occasionally quite far therefrom, for instance on the large island Hellas (40° S. Lat.), while it disappears from the surroundings and sometimes from the pole itself (the south pole). Such a highland covered by ice exists near the south pole, and is shown near the upper edge of Figure 24. In places where snow always remains, a feeble glacier formation may occur. Most investigators assume that mountains and plateaus exist on Mars, although of modest altitude (Campbell believes that he has observed peaks 3000 m. (9800 ft.) high). Lowell who diligently has looked for mountains at the edge of the illuminated part of Mars, has reached the conclusion that they, if present, cannot rise more than 600 to 900 m. (2000 to 3000 ft.) above the surrounding plains. It were indeed improbable that all inequalities of the Martian surface should have been removed in the process of disintegration, which although at work for enormous extensions of time, has long been extremely feeble and is unassisted by torrents of rain which might rapidly wash the products into the valleys. At present, it is mainly the sand carried by the desert wind that slowly reduces the roughnesses and in this process extensive highlands are hardly touched. But, without the assumption, in itself very unlikely, of a nearly level surface on Mars it becomes difficult to comprehend how the canals, if filled with pure water, can proceed in straight lines without reference to existing differences in altitude. Like the rivers on the Earth they ought to bend according to the topography, even if constructed by engineers.

When the canals freeze at the approach of winter, they invariably have been observed to disappear in company with the lakes or oases at their crossings. They are then all covered by the reddish-yellow dust carried by wind from the surroundings. When a canal is about to reappear it frequently first comes to sight as a dark streak evidently the result of moistened ferro-oxide. Occasionally a mist formation precedes the appearance of the canal. It is plain that the cold, misty air settles in the valleys, there as here, and gives up its moisture to the salts on their bottom and the canal is thus brought out as a dark line. Sometimes the vicinity also assumes a darker shade indicating the absorption of some moisture. On the sides of the canals the less hygroscopic salts are deposited. Possibly the green colour of the canals is partly a contrast-effect due to the red surrounding, possibly also the result of finely divided matter in the liquid. It is also conceivable that the cause is the reducing influence on the ferro-oxide of the sulphuric gases emerging from the fissures; an exceedingly small quantity accomplishes in this case large results. F. le Coultre describes the colour as being sometimes a dead black.

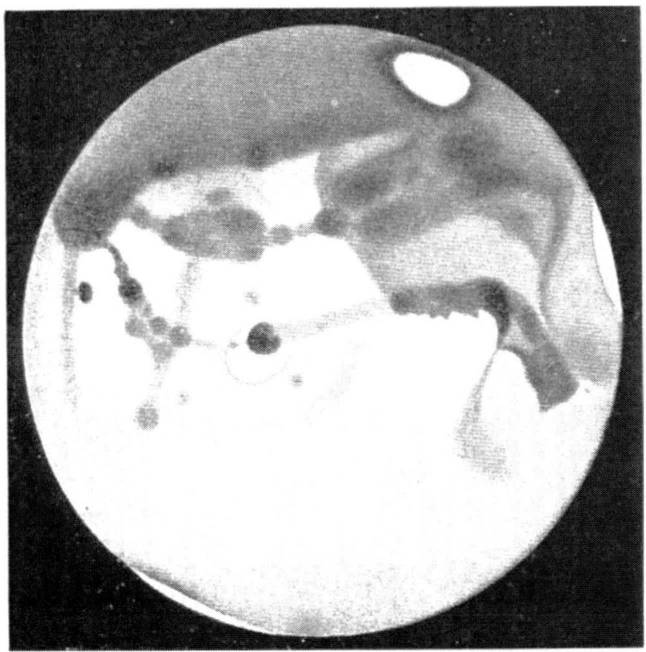

Figure 24. Mars, as observed by E. M. Antoniadi, October 6, 1909.

Something similar applies to the seas. When these freeze, especially in shallow places, yellowish-red dust from the continent settles on their surface and lends it hues between the original dark green and the light yellowish-red. When the ice subsequently melts this dust sinks in the water which latter resumes its dark green colour.

Chloride solutions, if concentrated, freeze at the following temperatures; that of calcium at -55° C. (-67° F.) that of magnesium at -44° C. (-48.2° F.) and that of sodium at -22° C. (-7.6° F.). If now, as we previously have seen, the mean temperature of Mars as a whole is about -40° C. (-40° F.), of the equatorial belt about -10° C. (+14° F.), and of the pole in the height of summer about 0° C. (32° F.), it is evident that a liquefaction of the ocean surface and of the canals, particularly where salts are deposited, very readily may take place. We should in this connection remember that the ice on Mars is stationary while on Earth it is in motion. The consequence is that sand and dust in the course of thousands of years have accumulated on the bottom of the shallow basins in the polar ice. These seas appear therefore dark in spite of their exceedingly small depth and the white salt and ice-crystals remaining undissolved are unable to display their light colour. Even in the "ocean," Lowell was persuaded that he had observed canals (see Figure 23), and it may be possible that cracks are in evidence there, particularly in the most shallow sections, as is the case in the Tyrrhenian sea north of Sicily. It is significant that Flammarion has reached the conclusion, which at first appears highly hazardous, that the freezing point of water is lower on Mars than on Earth. This is entirely correct, if we let water stand for salt solutions.

It is customary to point to the strictly uniform breadth and the rectilinear appearance of the canals as clear evidence of their being artificial, *i.e*, the work of engineers. The Italian astronomer Cerulli strongly objected to this conception. "In the exceedingly rare cases when both sides of the canal plainly may be seen," states Schiaparelli, "I have observed curves and notches in the borders." This occurred with the canals Euphrates and Triton in 1879, and with the Ganges in 1888. And it would seem obvious that watercourses produced in old furrows would not, as a rule, be of uniform breadth. Antoniadi, by his observations in the autumn of 1909 (see Figure 17a and 24), has confirmed this opinion, as has le Coultre, who found twice as many irregular canals as rectilinear ones. Antoniadi remarks that some canals appear to be collections of lakes strung out in a certain direction while others are narrow lines which bend and twist. "The complicated network of straight lines is probably illusory." The spots on Mars, he continues, are very irregular, and "present by no means any geometrical form" (on which the belief largely is founded that they are the product of intelligent beings). "The appearance of the planet reminds one of that of the Moon (except that the latter is dead, *i. e.*, unchangeable) or of a terrestrial landscape viewed from a balloon."—"In a word, the 'geometry' of Mars is revealed as a pure illusion." Exceedingly instructive is a comparison between the two maps of Mars drawn by Schiaparelli (1886) and by Antoniadi (1909) reproduced here and found at the end of the volume. While Schiaparelli as a rule represents the canals as narrow, straight, or slightly curved bands of uniform width, these formations on the Antoniadi chart frequently dissolve into a series of dark spots joined by less obscure sections (see for example the canals Nectar and Oeroe at the Sunlake). The same is true about several of the so-called "seas," particularly the Tyrrhenian (Mare Tyrrhenum), and the Sunlake (Lacus Solis); also about the "Ocean-bays" such as the well-known Syrtis major which with the Sunlake form the most conspicuous objects on the surface of Mars. These maps are, moreover, of great interest because several canals and other features present on one are absent on the other and vice versa. In this way, we obtain a vivid conception of the remarkable changeableness of the Martian surface as contrasted with the exterior of the Earth. The latter, if viewed from Mars, would not have presented any noticeable change in historical time except for the seasonal variation of the snow fields. This peculiarity of Mars is only explained by the fact that the geographical features of that planet as a rule are surface formations of a slight depth and therefore subject to rapid transformations.

Frequently, large white spots suddenly appear, especially near the lakes, such as the spot at Lake Phœnix near the centre of Fig. 24 which represents Mars on October 6, 1909, according to Antoniadi. These white spots disappear as suddenly as they show forth. The white colour is probably due to a very thin snow or hoar-frost, which is easily condensed in the vicinity of the lakes but which as readily vanishes at the approach of a warm draft or of sunshine.

Occasionally, dark spots on Mars are described as dissolving under strong enlargement into dark and light squares giving the appearance of a chessboard. This reminds one of the bayirs in Turkestan (see Figure 9).

The collections of lakes along the cracks on Mars which appear to us as "canals" are repeatedly filled up by sand and dried out. They are revived through new depressions along the dislocation fissures, corresponding to our earthquakes, when vapours of water and other gases pour forth and condense to lakes in the deepest pockets of the fissures. Canals are therefore created rather rapidly, sometimes overnight, and vanish occasionally as suddenly. The most remarkable case of "new" canals was made known through a communication by Lowell. Two new canals, at the time the most conspicuous on the surface of Mars, were observed east of "Syrtis magna" on September 30, 1909, from Flagstaff observatory, when they also were photographed, which precludes an illusion. (On the other hand there was no sign of the great canal Amenthes, shown on the map Figure 17, a short distance to the left, *ie*, east of Syrtis in the very section where the new canals were observed.) Also two new oases through which the new slightly curved canals passed were observed for the first time, as were also a few minor canals in the neighbourhood.

In 1913, the double canal Æthiops (see map at Long. 240°; the canal is there single) was rediscovered from the Lowell observatory after an absence of fifteen years.

These data make it evident that one or possibly several rather strong earthquakes took place east of Syrtis major just prior to September 30, 1909 with the two oases as centres of collapse. The fissures now made visible have probably existed before but filled with sand and have now reappeared as a result of the condensation of water vapour when it emerged into the cold Martian air.

This fact, that the most prominent canals in such manner now suddenly appear and now as rapidly vanish, ought to convince us beyond doubt that they are not magnificent products of engineering skill, for the construction of which we should require centuries on the Earth.

The theory that intelligent men exist on Mars is very popular. With its help everything may be explained, particularly if we attribute an intelligence vastly superior to our own to these beings, so that we not always are able to fathom the wisdom with which their canals are constructed. The crossings of the latter are said to be cities (Lowell) fifty times greater than London. The trouble with these "explanations" is that they explain anything, and therefore in fact nothing. If we would endeavour to understand the phenomena on Mars, we must in the first place avoid the formerly so popular principle of "purposiveness" which led even the most prominent scientists into so many amusing errors. Neither may we base our conceptions, as does Flammarion, on the assumption of natural forces unknown to us, no matter how much such a course may appeal to mystics. Only forces with which we are familiar can be resorted to, if we really are to understand nature. It seems to me that such method of research might with good results be applied also to the planet Mars.

Chapter 7

MERCURY, THE MOON, AND VENUS

The planet Mercury probably resembles Mars in many respects, but differs particularly in lacking an atmosphere. The fissures in the crust of the Earth or of Mars are as a rule rapidly filled and their contours largely hidden from sight by alluvium or sand carried by windstorms, so that they reveal themselves only through tremors and various emanations along their course. The fissures on Mercury on the other hand must remain as yawning chasms. It is probable that reducing gases stream out of these cracks as on Earth, and colour the environment in a darker shade than the other visible part of the planet's surface, that is the hemisphere turned toward the Sun. Not very volatile gases, such as sal-ammoniac, other chlorides and sulphur, which on the Earth are deposited inside the fissures, may here spread over large areas and discolour surrounding territory, particularly where iron compounds are present, and, under the attack of sulphur, turn black. Lowell has made drawings of the dark spots visible on Mercury, one of which is reproduced as Figure 25. These spots lie, as on Antoniadi's drawing of the surface of Mars (Figure 17a), arranged in lines which are almost straight or of a slight curvature only. This seems to indicate that the spots belong to areas immediately adjoining enormous fissures. According to Lowell's drawing, these cracks are far more regularly distributed on Mercury than on Mars. Very close to the centre of the ever-sunny side we see a dark spot, a "lake." It is evident that this spot is located in by far the hottest point on the surface of Mercury. This gives rise to the following conception. The hottest part of Mercury was naturally the last to solidify. Mercury evidently ceased to rotate around its own axis, leaving one side continually exposed to the Sun, while its surface yet consisted of lava that was fluid, at least where the sunshine was most intense. The weakest point on the planet was therefore the one just opposite the Sun. When later collapses occurred the cracks commenced at this weak point. We see on the figure how not less than six fissures radiate from this centre. Others were formed where the crust broke off from adjacent solid portions. These latter fissures have a less rectilinear appearance than those diverging from the centre of collapse. Along these faults, reducing gases no doubt issue from the interior of the planet and give a dark tone to the surface layers, which probably consist of ferruginous dust falling from space. In the neighbourhood of the Sun, such dust ought to be more plentiful, concentrated as it were by the gravitation of the Sun. Mercury lies five times nearer the Sun than the Earth does, and twelve times nearer than Mars. There probably also exist on Mercury, as on the Moon, large mountains which are not subject to the wear of running water and blowing sand. We cannot, however, observe them from the Earth. Possibly they correspond to the widely extended spots, which several investigators as Schroeter, Vogel, and others, have noticed, formations resembling the "seas" on the Moon. Vogel believed that he had found traces of water vapour in the atmosphere of Mercury as in that of Mars, a belief in both cases undoubtedly founded on erroneous observations.

The part of Mercury which is turned from the Sun must be characterized by a tremendous cold due to radiation into space. The temperature stays probably about 200° C. (360° F.) below the freezing point of water (328° below zero F.). Even the most concentrated solutions we know of freeze to ice precipitating the salt considerably above this temperature. Moisture in fluid state can, therefore, not very well exist on this side. On the sunny hemisphere it must be lacking as well, due to evaporation over to the cold side. As a result, the desolation on Mercury must be far greater than that on Mars and surface changes caused by variations in temperature are almost precluded. On account of the so-called libration, certain dark portions near the boundary of the illuminated hemisphere occasionally enter the sunlight. But, during this interval, all traces of moisture are undoubtedly driven away from these parts, also never to return.

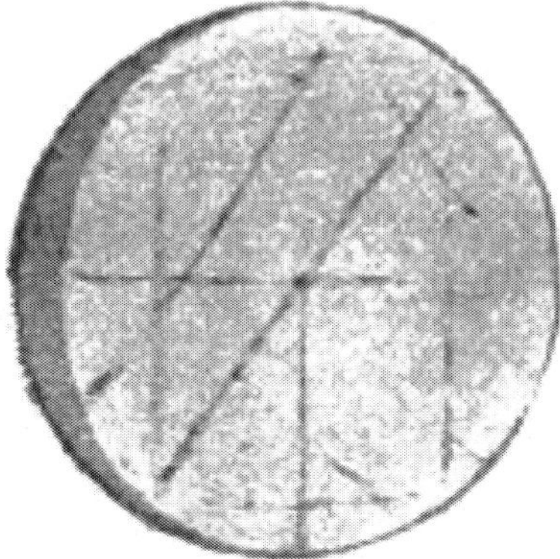

Figure 25. Drawing by Lowell, representing the planet Mercury with "canals."

Figure 26. A part of the moon near its south pole. The big crater above, in the interior and on the walls of which a large number of smaller craters appear, is Clavius. A little below and to the right is Longomontanus just at the edge of the shadow; almost in the middle of the picture appears Tycho with its central cone. The moon diameter corresponds to 43.4 cm. Photo by Yerkes Observatory.

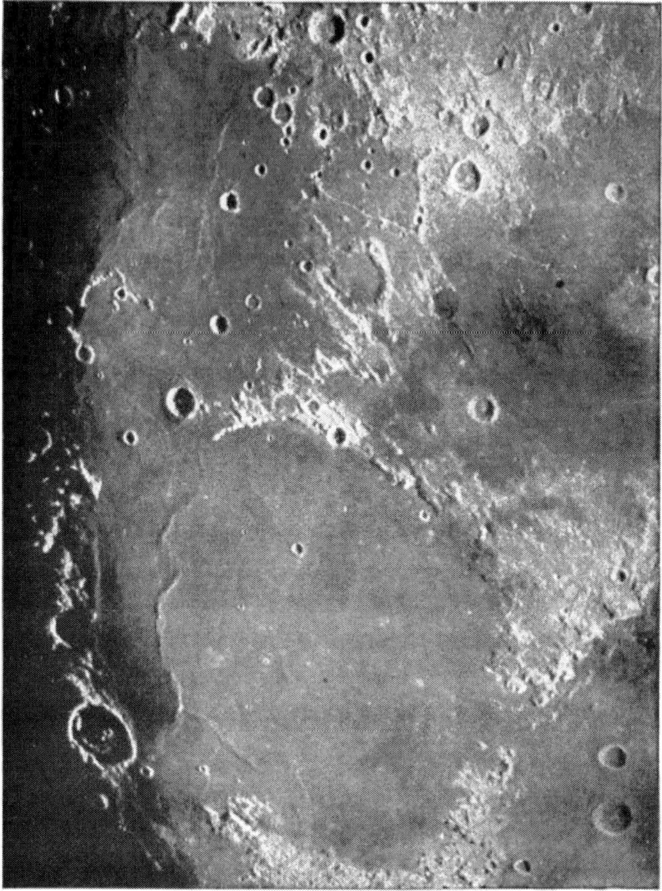

Figure 27. Mare Serenitatis (below), Mare Tranquillitatis (upper left) and vicinity. To the left of Mare Serenitatis the great crater Posidonius; 2.8 cm. from the right edge a small white spot may be seen. This is the remarkable crater Linné, said to have undergone changes. The moon diameter corresponds to 35.7 cm. Photo by Yerkes Observatory.

The Earth's moon is not entirely as stagnant as Mercury, although on the whole it closely resembles this planet. The Moon always turns the same side to the Earth—a small libration exists here also—so that each part of its surface is illumined by the Sun during one half of the synodical month (29.53 days). This time, however, is so long that the moon's surface in the meantime almost assumes the temperatures due to continuous sunlight and continuous night.

Some investigators, as W. H. Pickering, are persuaded that portions of the moon just emerged from the shadow show a lighter colour than after a short time of illumination. These observations, however, have not been accepted as correct. According to Pickering, the light colour should result from a slight formation of snow or hoar-frost during the long night of 355 hours. If an appreciable trace of vapour existed on the Moon, it ought to evaporate and form white caps over the poles where the Sun's heat is not sufficiently strong to melt them. As no such signs have ever been observed, the faith in snow on the Moon is not likely to find many defenders.

The lunar mountains are not attacked by water or sandstorms, nor do they peel off due to rapid heating by the Sun. They rise, therefore, to full stature over their surroundings. Their height can be measured by the length of their shadows. Mädler, in this manner, computed one of the peaks of the Mountain Newton to rise 7300 m. (24,000 ft.) above the territory on which its shadow falls. Six peaks reach between 6000 and 7000 m. (19,500 and 24,000 ft.), 21 between 5000 and 6000 m. (16,500 and 19,500 ft.), 82 between 4000 and 5000 m. (13,000 and 16,500 ft.), and 582 reach 2000 m. (6500 ft.) and more. These figures show the extraordinary mountainous character of the Moon's surface compared to that of the Earth which is thirteen times larger.

In Figure 26, we see a picture of the portion of the Moon most rich in volcanoes, with the crater Tycho in the centre and Clavius above.

The numerous volcanoes are particularly characteristic of the Moon. They vary in magnitude from a diameter of over 200 km. (125 miles), for example the colossal Clavius with its companion craters, down to dimensions just visible with the aid of a telescope. The largest exceed our biggest many times in width and differ essentially from them, inasmuch as their bottoms are flat, occasionally provided with smaller volcanic cones—see the crater Longomontanus to the right of Tycho on Figure 26—and surrounded by a high (inwardly often very steep, outwardly more sloping) wall as on Clavius, Longomontanus, and Tycho. The largest, as for example Clavius, may be compared to a province such as Bohemia, surrounded as it is on all sides by mountains. The elevated ring, as well as the interior of Clavius, is adorned with numerous large and small craters.

The smallest of these resemble hemispherical excavations in the crust of the Moon, or they may be small volcanic cones which break through the walls. Sometimes they are strung out like pearls along rents in the ground.

All these volcanoes have undoubtedly given passage from the interior to the surface of the Moon for enormous volumes of gases previously enclosed in the lunar magma. Nor is it less certain that these gases have consisted largely of water vapour. If this had been condensed to water, oceans and rivers would have been formed, and on the bottom of the seas would have been deposited sediments carried down from the mountains. Such, however, is not the case. The so-called "seas" on the Moon are indeed on a lower level than their surroundings, but their surface is even (see Figure 27 with Mare Serenitatis below and Mare Tranquillitatis above to the left; see also Figure 29 with Mare Imbrium below; it is bounded on the right by the "Carpathians"). The "seas" consist of volcanic rocks, and are not at all covered with loose sediments which if present ought to reflect light better than the volcanic vitreous rocks. But the lunar "seas" are much darker than the environments. This shows that seas proper, or bodies of water, have probably never existed on the Moon. Even before the surface had changed from its molten condition the water vapour had departed from the atmosphere, and the new quantities which the volcanoes emitted from the depths below disappeared so rapidly that lakes were never formed. The history of other atmospheric gases on the Moon was no doubt similar. All evidence, therefore, points to the conclusion that life never inhabited its rough surface. Figure 27 shows that the "sea-bottoms" are not free from volcanoes. They also abound in folds, corresponding to mountain-chains on the Earth. These folds indicate old breaks in the crust while it was yet very thin. To the right, in Mare Serenitatis, appear a few white spots which W. H. Pickering ascribed to snow. The largest is the much discussed "crater" (?) Linné. Mare Serenitatis is surrounded by a ring of volcanoes.

A noted astronomer, Cerulli, observed, when he directed a glass of moderate power, such as opera glasses, toward the Moon that the spots seemingly arranged themselves in rows forming intersecting lines similar to the canal-system on Mars. As the regularity disappeared with greater enlargement, Cerulli believed that the canal-system on Mars also would dissolve into small spots if a sufficiently powerful telescope were used. His idea, which partly has been verified, was more recently adopted by the Englishman, Maunder, who denies the existence of canals on Mars. Photography, however, has proved their reality (Figure 18).

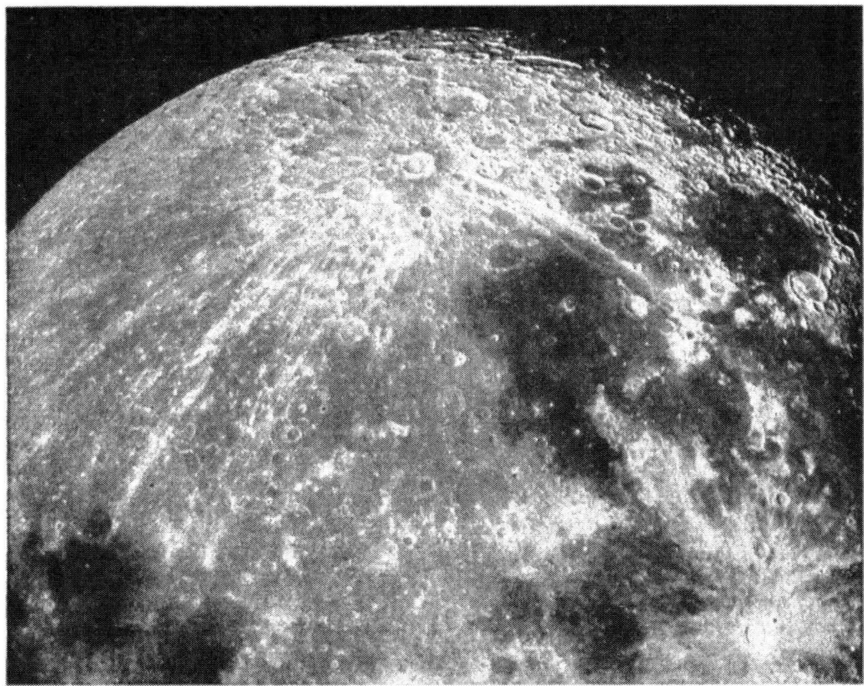

Figure 28. Tycho in full illumination with surrounding magnificent system of streaks. In the lower right corner Copernicus with a less regular system appears. Between them Mare Nubium, in the upper right Mare Humonum, with the great crater Gassendi below. The moon diameter corresponds to 16.7 cm. Photo by Yerkes Observatory. Compare Figures. 25 and 28, showing parts of the same territory under side light.

If we disregard the illusory reticulation, there are nevertheless on the surface of the Moon numerous designs of a nearly rectilinear outline. There are to begin with the sinuses, extended trenches, often dotted along their sides with minor volcanoes. Figure 27 shows, in the upper right corner, two such sinuses, the right one with a small volcano, Hyginus, in the middle. There are, further, five such volcanoes in its left arm, not visible on the photograph, and two in the right arm. The second, "Sinus

Ariadaeus," commences to the left with the Volcano Ariadaeus, not visible on the figure. The explanation of the origin of these sinuses is probably to be found in the different contraction of the Moon's surface layer and of the hotter substrata immediately after the solid crust was formed. In a way, they correspond therefore to cracks in the glazing on porcelain. Like the two sinuses just mentioned they frequently commence and end with small craters which formed weak spots in the crust that facilitated the original break. Later on, volcanoes broke through along the sinuses themselves. In several regions of the Moon, and particularly in the equatorial belt, observers have claimed discoveries of new sinuses and occasionally of minor craters, "which could not possibly have escaped notice if they had existed before." At present, the almost unanimous verdict is that such changes are very improbable, and that the visibility of the "new" objects largely depends on favourable sidelight, so that they might well have been overlooked if the region in question previously was examined under less advantageous illumination.

The most peculiar formations on the Moon are the so-called "bright streaks" which as a rule issue in almost straight lines from some of the larger craters, particularly Tycho and Copernicus. Those around Tycho (see Figure 28) do not seem to be neither raised above nor depressed below the surroundings to any degree worth mentioning. For this reason they are not visible under oblique illumination as in Figure 25. They proceed in straight lines independent of elevations. This quality is in striking conformity with the characteristics of fissures on earth as for instance those that traverse the Tyrrhenian Sea and the Calabrian mountains. They also resemble the canals on Mars in this respect. Nasmyth and Carpenter caused a glass ball, containing water under pressure, to break at one point and obtained a system of beams radiating from this point and vividly reminding of the streaks around the lunar craters. The same effect appears if a homogeneous plate, of glass for instance, is broken by a blow in one point. No doubt these streak centres were once centres of collapse although they sometimes now are found at a considerable elevation, like Tycho. This may be the result of a later secular lifting of the rocky substrata like the slow rise of the Scandinavian Peninsula. The streaks around Copernicus (see Figure 29) are very different from those around Tycho. They are not rectilinear and consist next to the crater of distinct mountain chains plainly visible under oblique illumination. They penetrate into Mare Imbrium (Figure 29 below) crossing the mighty "Carpathian" mountains. Frequently, they are provided with minor volcanoes, as in the streak directed almost straight downward on the figure, *ie*, to the north. They are obviously volcanic fissures like those on the Earth.

Figure 29. The great lunar crater Copernicus surrounded with streaks. Below the Carpathian mountain range and at bottom part of Mare Imbrium. The moon diameter corresponds to 55 cm. Photo by Yerkes Observatory.

The streaks, in many cases, would not be visible at all were it not for their different colour, which is considerably lighter than that of the surroundings. The only explanation offered for this fact is the assumption that the original cracks were filled by some light matter forced out from the interior of the Moon, that is by the lunar magma. This magma was not very viscous, as it has spread out considerably beyond the edges of the cracks proper. These presumably, like those on Earth, were of a rather moderate width, not enough to be distinguishable at the Moon's distance. Similar light-flowing emanations from long fissures are known also on our planet, for instance, from the Laki eruption on Iceland in 1783. The colour may be light simply by comparison with the previously solidified crust, which, optically, as regards reflexion of light, has proved very similar to obsidian or, even more like another volcanic mineral product, vitrophyre. It is also possible, however, that gas bubbles were liberated as the lava solidified and gave the surface a milkwhite appearance—gravity on the Moon is only one sixth of that on the Earth so that the bubbles would rise and evaporate extremely slowly from the magma. Due to the very low atmospheric pressure on the Moon, the bubbles would also occupy a larger volume than in a corresponding case on Earth and become more conspicuous in proportion. They probably partly remained on the surface of the outpoured lava as a thin scum, which hardened in that state. Since then, it has suffered no more change than all other formations on the Moon, while on the Earth it would soon have been scoured away by sand and water.

Before we leave the Moon it may be well to say a few words about its colour. Mädler states in agreement with several other observers that Mare Serenitatis, a "sea" on the Moon's north side (25° latitude) just to the right of the centre meridian (see Figure 27) is remarkable for its beautiful pure green colour, while Mare Crisium about 16° Lat. N. near the right edge of the Moon is of a dark grey-green hue. In Mare Humorum (about 22° Lat. S., not far from the edge of the Moon, see Figure 28) grey and dark green shades alternate and in Mare Frigoris, just inside the lunar north pole, the colour is a dingy yellowish-green. In other words, the characteristic colour of the great lava seas is apparently green. This agrees closely with conditions on the Earth where similar formations are coloured green by silicates of ferrous protoxides, certain species of which are called green-stones. Franz, however, questions the observations of Mädler and professes the belief that very light craters appear bluish and assumes this to be a contrast effect to the general yellow hue of the Moon. Langley investigated the lunar radiation with the spectroscope and found that the ratio of blue to yellow was smaller in the moonlight than in the sunlight, for which reason the general colour of the moon resembles that of yellow sandstone.

A very interesting observation was made at the Lowell observatory when investigating the spectrum of the sparse light reflected from the Earth to the portions of the Moon not exposed to the sunlight. It proved to be of a far more blue tinge than sunlight reflected from the Moon. Our conclusion must be that the Earth shines with a blue lustre. This is perfectly natural, as the diffused light which reaches us after having been scattered by particles suspended in the air (and by gas molecules as well) is a deep blue and there exists no reason why that part of the light which is thrown outward into space should be of a different colour. The Earth, therefore, is blue in contradistinction to Mars which is red, on account of its desert surface, and Venus which is bright white. The cloudy portions around the equator and the poles should appear light blue from without and should be separated by dark blue bands over the so-called horse latitudes, under which the cloudless desert regions are located on either side of the equator.

Compared to Mars the Moon offers a scene of far greater desolation. On Mars, we observe at least some considerable changes such as the disappearance of the white pole-caps at midsummer when at the same time a dark ring appears to surround them; then the "lakes" and the "canals" come into view, beginning close to the ring mentioned, later on nearer the equator, and finally on its other side, while the opposite pole-cap puts on its winter hue. Again, we have the sudden appearance, and equally hasty disappearance, of white spots, particularly in the neighbourhood of the lakes, and the sand storms which hide the surface of Mars and often fill its canals. The abruptness of the changes indicate that they are confined to a very thin surface layer. The formation, on the other hand, of canals, for many years unobserved, must be ascribed to a volcanic activity which, while feeble, yet must be seated in the deeper portions of the planet. In addition, a stunted vegetation of low forms is not unthinkable in the polar regions.

As against this, the Moon is undoubtedly a stellar body entirely insusceptible of surface change. Near its centre, it is probably not completely solidified and an extremely slow growth of the firm crust is therefore likely. Gases are no doubt set free during this process, but they are unable to penetrate the enclosing thick armour and remain therefore as bubbles in the hardening magma.

As a matter of fact, no changes have, with certainty, been detected on the Moon's surface. It is true that the great Wm. Herschel, known as an excellent observer, believed that he discovered, in 1873, mountains which had not existed before that time, and Schröter, who diligently studied the lunar surface, was of the opinion that he too had discerned numerous changes. These discoveries, however, were doubted by careful critics, and after the publication of Mädler's great work about the Moon (1837) the complete stagnation on that body was taken for granted. Nevertheless, there are several astronomers, such as Schmidt in Athens (1866) and lately W. H. Pickering in Cambridge, Mass., who think that they have discerned considerable modifications. The former held that the crater Linné (Figure 27) had vanished since the publication of Mädler's work. In 1867, Mädler, himself, proclaimed that it had the same appearance as before. Pickering, again, reports periodic changes of "snow" and "vegetation." (Compare Figure 27 taken from Pickering's Moon-atlas.) Closer analysis, however, indicates that the phenomena are probably only apparent and depend on the angle of illumination at each particular time of observation. For some time, rather more than a quarter of a century, photography has been pressed into the service of lunar investigation with far more objective results than would be possible through direct ocular inspection alone. During this period, which, it must be admitted, is not very long, no distinct signs of changes have been recorded by the photographic plates.

The great difference between Mars and the Moon depends upon the existence of a real atmosphere on the former. The oxygen will probably vanish from Mars also, being used up in the course of disintegration. But nitrogen, argon, and the other permanent gases will always remain, as will the water vapour from the bodies of water ever present, particularly around the south pole. It is true that this water vapour also will diminish with sinking temperature and when the latter finally has reached the freezing point of the salt solutions on Mars, the canals and the lakes will cease to thaw out or liquify under the vapour distilled over from the warm to the cold pole. But sand storms and thin mist formations will always appear and cause colour changes on the desolate planet.

If we wish to picture to us the future fate of our Earth when it gradually enters the reign of darkness and cold in consequence of the enfeebling of the Sun, we must seek our illustration on Mars and not on the Moon. Slowly are the oceans going to freeze, finally down to their bottom, the abundance of the rainfalls will diminish, only light snow will now and then bring change to a surface evermore transformed into a sand desert as far as the continents reach. Rents in the rocky substrata of the latter will appear as dark lines, caused by the gases rising from the interior. When the temperature at the equator has fallen below the freezing point, the polar regions will remain the only parts where a light covering of frost will melt in the height of the summer season and where the last feeble organisms will eke out their hard existence, resorting to a prolonged winter's sleep of their seeds and spores. Finally, the last remnant of life will also disappear and sandstorms alone, save for the gasps of gas emanation from fissures in the rocky ground, will bring relief to the monotonous desolation. Falling meteoric dust, which now exists in original state only on the bottom of the oceans, will gradually cover the entire surface of the Earth with a mantle coloured brick-red through the influence of atmospheric oxygen. When the oxygen itself is used up, the meteoric dust will retain its original greyish-green hue and lend it to the funeral pall of the Earth.

Very different conditions obtain on our neighbour planet, which is closer both to the Sun and to ourselves, the radiant Venus, an object of interested human attention already in ancient times. The average temperature there is calculated to about 47° C. (116.6° F.) assuming the sun constant to two calories per cubic centimeter (.061 cu. in.) per minute. The humidity is probably about six times the average of that on the Earth, or three times that in Congo where the average temperature is 26° C. (78.8° F). The atmosphere of Venus holds about as much water vapour 5 km. (3.1 miles) *above* the surface as does the atmosphere of the Earth *at* the surface. We must therefore conclude that everything on Venus is dripping wet. The rainstorms on the other hand do not necessarily bring greater precipitation than with us. The cloud-formation is enormous and dense rainclouds travel as high up as 10 km. (6.2 miles). The heat from the Sun does not attack the ground but the dense clouds, causing a powerful external circulation of air which carries the vapour to higher strata where it condenses into new clouds. Thus, an effective barrier is formed against horizontal air currents in the great expanses below. At the surface of Venus, therefore, there exists a complete absence of wind both vertically, as the Sun's radiation is absorbed by the ever present clouds above, and horizontally due to friction. Disintegration takes place with enormous rapidity, probably about eight times as fast as on the Earth, and the violent rains carry the products speedily downhill where they fill the valleys and the oceans in front of all river mouths.

A very great part of the surface of Venus is no doubt covered with swamps, corresponding to those on the Earth in which the coal deposits were formed, except that they are about 30° C. (54° F.) warmer. No dust is lifted high into the air to lend it a distinct colour; only the dazzling white reflex from the clouds reaches the outside space and gives the planet its remarkable, brilliantly white, lustre. The powerful air currents in the highest strata of the atmosphere equalize the temperature difference between poles and equator almost completely so that a uniform climate exists all over the planet analogous to conditions on the Earth during its hottest periods.

The temperature on Venus is not so high as to prevent a luxuriant vegetation. The constantly uniform climatic conditions which exist everywhere result in an entire absence of adaptation to changing exterior conditions. Only low forms of life are therefore represented, mostly no doubt belonging to the vegetable kingdom; and the organisms are nearly of the same kind all over the planet. The vegetative processes are greatly accelerated by the high temperature. Therefore, the lifetime of the organisms is probably short. Their dead bodies, decaying rapidly, if lying in the open air, fill it with stifling gases; if embedded in the slime carried down by the rivers, they speedily turn into small lumps of coal, which, later, under the pressure of new layers combined with high temperature, become particles of graphite. Fossils proper are not formed as was also the case in the early periods of the Earth.

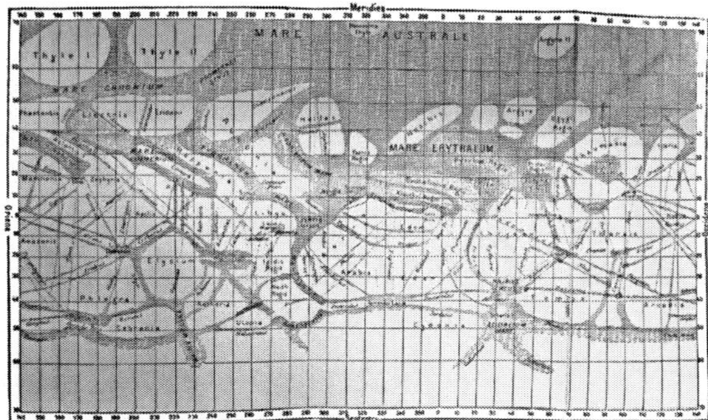

Figure 17. Map of the planet Mars in Mercator's projection according to drawing by Schiaparelli. A comparison with the drawing by Antoniadi (Figure 17a) suggests that Schiaparelli has made a somewhat diagrammatic picture, which sets forth a very great number of strictly straight "canals."

(larger)

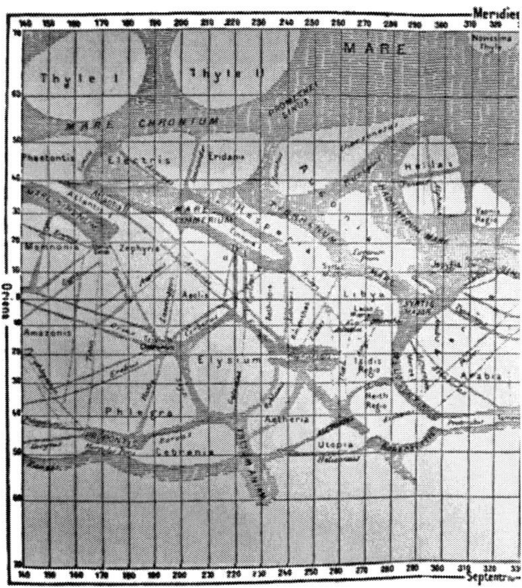

Figure17a. Map of the planet Mars in Mercator's projection, drawn in 1909 by E. M. Antoniadi.

The temperature at the poles of Venus is probably somewhat lower, perhaps about 10° C. (18° F.) than the average temperature on the planet. The organisms there should have developed into higher forms than elsewhere, and progress and culture, if we may so express it, will gradually spread from the poles toward the equator. Later, the temperature will sink, the dense clouds and the gloom disperse, and some time, perhaps not before life on the Earth has reverted to its simpler forms or has even become extinct, a flora and a fauna will appear, similar in kind to those that now delight our human eye, and Venus will then indeed be the "Heavenly Queen" of Babylonian fame, not because of her radiant lustre alone, but as the dwelling place of the highest beings in our solar system.

The ancients believed that the fates of men could be read in the stars and this faith persisted with the power of a religion until a few centuries ago. It was shared by the foremost astronomers, pre-eminently by Tycho Brahe, who endeavoured to support it through his investigations. Traces are yet to be found in popular conceptions. These ideas have been verified today in a certain sense although with a wholly different meaning than held by our forefathers. The planets do tell us the conditions that existed on the Earth at the first dawn of life and we can also draw from them a prediction of the fate that once, after milliards of years perhaps, will befall the latter descendants of present generations.

In one respect the dreams of our ancestors have not proved true, namely, with reference to the habitability of the other globes in our solar system. According to the great Kant, conditions on the wandering stars outside of the Earth's orbit were so favourable to life that their inhabitants ought to have reached a far higher development than beings on the Earth. The last remnant of this conception lives in the speculations about the marvellously proficient engineers who built the magnificent system of giant canals on Mars. A thorough critique has demonstrated that any other planet in our solar system hardly can offer an abode for higher beings, except this very Earth, which therefore justly may be called "the best of worlds" among those that we know. And yet, it was undoubtedly a great truth that Giordano Bruno gave his life for, because it is highly probable, nay almost certain, that around the countless suns which dot the firmament spin dark bodies, although unfortunately our most powerful lenses do not reveal them. A number of these unseen stellar bodies shelter living beings, which even might have climbed to a higher point on the ladder of evolution than have the inhabitants of the Earth.

Explanations: Meredies = south; Oriens = east; Occidens = west;
Septentrio = north; Nix = snow.

Abbreviations: M = Mare, sea; S = Sinus, bay; Fr = Fretum, channel;
L = Lacus, lake; Fl = Flumen, river;
R = Regio, region; I = Insula, island;
Pr = Promontorium, cape.

(larger)

TRANSCRIBER'S NOTES

Punctuation and spelling were made consistent when a predominant preference was found in this book; otherwise they were not changed.

Simple typographical errors were corrected; occasional unbalanced quotation marks retained.

The spelling and accent marks of non-English words have not been changed.

Ambiguous hyphens at the ends of lines were retained; occurrences of inconsistent hyphenation have not been changed.

Illustrations have been moved, when necessary, to fall between paragraphs and outside of quoted text. In some cases, this places them on different pages than the ones given in the Table of Illustrations. In versions of this eBook that support "links", the page numbers in that table link to the correct illustrations.

INDEX

A

B

C

D

E

G

H

I

K

L

M

N

O

P

S

T

V

W

Z

Neutron Stars: The Aspect of High Density Matter, Equations of State and Observables

Editors: Hiroshi Uechi (Osaka Gakuin University, Osaka, Japan), Schun T. Uechi (Osaka University, Osaka, Japan) , Brian D. Serot (Indiana University, Bloomington, Indiana, USA)

Series: Space Science, Exploration and Policies

Book Description: This book illustrates observed values and properties of neutron stars, field theoretical approaches of hadrons to high density matter, such as nuclear and hyperonic matter, signals of dynamical phenomena, hadronic and hadron-quark neutron stars, radiations from high density matter and supernova phenomena.

Hardcover ISBN: 978-1-62081-123-8
Retail Price: $195

Neutron Star Crust

Editors: Carlos Bertulani (Texas A&M University Commerce Commerce, TX), Jorge Piekarewicz (Florida State University, FL)

Series: Space Science, Exploration and Policies

Book Description: Neutron stars are gold mines for the study of nuclear systems under extreme conditions of density and isospin asymmetry. Spanning many orders of magnitude in density and pressure, neutron stars display exotic phases that cannot be realized under normal laboratory conditions.

Hardcover ISBN: 978-1-62081-902-9
Retail Price: $210

Telescopic Work for Starlight Evenings

Author: William F. Denning

Series: Space Science, Exploration and Policies

Book Description: This book was written to be intelligible to general readers while also containing facts and figures useful to amateur astronomers at the time of its first printing in 1891.

Hardcover ISBN: 978-1-53614-230-3
Retail Price: $230

Next Generation of Human Space Flight Systems

Editor: Alfred T. Chesley

Series: Space Science, Exploration and Policies

Book Description: This book is dedicated to important issues concerning Space Shuttle retirement and transition to the next generation of human space flight systems. Emphasis is placed on the issues of costs, suppliers, transition progress, environmental risks, and mission identification and scope.

Hardcover ISBN: 978-1-60692-726-7
Retail Price: $130